RED SKY

A.B. Acharya

For Cheryl, Devi, and Maya

Part I

One

July 18, 2006

The interrogation room looks exactly as I imagined. The walls are made of concrete blocks painted dull yellow. The fluorescent light flickers overhead. I sit on a stiff plastic chair, pen in hand, writing this. In front of me is a stack of white paper, which I believe is standard copy paper. The metal table in the middle of the room is bolted to the floor and takes up most of the space.

I'm alone now. Officer Coffin was here with me just a few minutes ago. He said he wanted to know everything about my relationship with Ian and Sophie. He wanted the whole story, or more precisely, my confession. He handed me three ballpoint pens and a stack of paper and told me to get started.

He's a big guy—Officer Coffin—with a thick walrus mustache. His voice is a deep rumble, and his movements seem somewhat agitated. He strikes me as the kind of guy who's always overworked and constantly pissed off. And I've just added a lot more work for him. I totally understand. It's exhausting to think about how much work poor Officer Coffin must do because of me. If I had died, it would have been easier for everyone involved, including myself.

Where do I even start? Living through these past few months, I never saw it as a story with a clear arc or anything like that. Now that I have to write it all down, I need to explain it in a way that makes sense. It's ironic because it didn't make sense at the time. It's good that Red Sky gives you a mental boost and helps you see the big picture.

The irony of all ironies is that Red Sky, which almost killed me, now provides me with the calm and clarity to explain how I got through all the shit that just happened. I'm not a professional writer, so I appreciate your understanding in advance. I'll do my best to make this as accurate and informative as I can. Okay, here we go.

My name is Narin Roy. I'm thirty years old, born in 1976. I grew up in the suburbs of Chicago as the son of Indian immigrants, Joyti Roy (my dad) and Mina Roy (my mom). My parents brought me here when I was three years old. My sister, Deepa, is seven years younger and was born in the States. Deepa just started medical school. She's a good kid. Although we're very different, we mostly get along well. She and my parents are probably all worried about me. Next time Officer Coffin sticks his head in, I'll ask if I can call my family.

Let's begin the story on the day we held a memorial service for my grandmother, my dad's mom. I was in my hometown of Chicago at the time. I'd just finished my postdoc a month earlier and was mentally preparing for the most important interview of my life. On that day, April 13, 2006, I met Ian and Maru for the first time.

I stood in the foyer of the funeral home's main hall, facing a large black-and-white photo of my grandmother. The glossy picture, mounted on a metal easel, reminded guests of who they were there to mourn. I liked the photo. Her wispy gray hair shimmered in the soft light, and that distinctive mole cast a shadow over her left brow. Her eyes sparkled, while her crooked teeth and uneven lips formed a mischievous smile.

I held a garland of white flowers. Mom had handed it to me and told me to drape it over my grandmother's picture. The garland was heavier than I expected. The flowers were thick and fleshy, but they felt nice in my palms—cool, soft, and silky. They carried a slight floral scent, which wasn't overwhelming or offensive. I wrapped the garland securely around the top edges of the frame and let it fall into place.

Stepping back, I examined my handiwork. My reflection was superimposed on my grandmother's face. I felt (not for the first time) that we shared no resemblance or connection. I tried to feel sad. A human had died. And she wasn't just anyone, but my only remaining grandmother. I was almost there. I had to think of dead kittens to get the sadness flowing. But then someone tapped me on the back, smiled, and walked away. He was one of my parents' many friends. The gesture was meant to be friendly, but it instantly reminded me of my farewell party, and I fell into the familiar, terrible rage.

The more I tried to suppress my anger, the more it churned and whirled inside me like a vicious, poisonous snake. My mind was filled with painful images—Jerome Cohen slapping me on the back, congratulating me on finishing my postdoc, and wishing me luck. The day of my farewell party should have been a victory lap, but Cohen turned it into a walk of shame. I hadn't shared my feelings with Dr. Cohen that day. I just nodded politely and smiled. I'd been working on containing all my negative emotions, just as my therapist instructed.

The funeral hall was soon filled with a low hum of idle chatter and the scent of flowers, perfume, and incense. I watched my dad talking with the round-bellied priest, who was dressed in a *dhoti*—a white cloth wrapped around his stout frame. The priest nodded solemnly to my dad before turning his attention back to an oil lamp beside a display of flowers and fruit. He began to chant as he arranged everything in place. I was the only one paying attention to him.

I closed my eyes and tried to focus on the Sanskrit words muttered by the priest. I had no idea what he was saying, but the cadence was soothing. Frustrated with myself and the agitation that had overtaken me, I slipped back into my usual meditation on the stupidity of emotions. They roiled beneath the surface, hidden from view, beyond our control, yet they govern us in ways we don't even realize.

Humans have tamed much of nature, but we still can't eliminate the negative feelings that torment us: sadness, envy, loneliness, and hate. Still, I looked forward to the day when this major problem would be solved. If

I succeeded in my work, no one in the future would have to suffer from the gut-wrenching hate and terrible uncertainty that held me prisoner at that moment.

I saw my sister, Deepa, taping colorful, glossy pictures of the deities on the wall. The bright images depicted gods and goddesses wielding powerful weapons: some smiting wicked demons, while others smiled in divine bliss. Deepa tugged awkwardly at the brightly colored *salwar kameez* she was wearing. As it slipped from her shoulder, she grabbed the voluminous shawl and threw it to the floor in frustration. I chuckled at her annoyance and walked over to her.

My sister is a few inches shorter than I am, about 5 feet 3 inches. She has a mop of black, unruly hair that she used to spend hours combing every day in junior high. She's blossomed into a pretty girl, but you didn't hear that from me. She's not gorgeous, but she's cute. Unlike me, she has lots of friends, yet she struggles to attract the men she likes. Her earnestness makes guys see her as a friend rather than a girlfriend.

"Hey, Deepa, do you need any help?"

"I need this to be over," she hissed.

The strong scent of perfume and flowers drifted by, and I felt a slight pressure building inside my head. I reached into my pocket, tore open my migraine rescue medicine, and swallowed it in one quick gulp. No pain yet, just pressure. It might stay that way. If not, I knew I was in for a long afternoon.

Deepa sank into a folding chair, and I took a seat beside her. "Do you remember going to India?" she asked. "When was that? 1986? Has it been twenty years?"

"I remember a few things," I said. "I was ten, so you would've been three."

"No wonder I can't remember anything."

"You don't remember her from that trip?" I nodded at our grandmother's picture. "She was always sneaking sweets to us when Mom wasn't looking."

Deepa shook her head.

"We played Snakes and Ladders with her every day. You'd get so upset if you didn't win. Remember that?" That summer in India, I spent hours finger-tracing the paths of those colorful, beady-eyed, fork-tongued snakes. I was fascinated by the small wooden dice; I'd never seen dice made of wood, only plastic. It would be wonderful to have a game board like the one we played in India, with colorful pieces, wooden dice, and snakes slithering in a vibrant world all their own.

"No, I don't remember anything," Deepa replied mournfully.

"While we played, she told us stories about the gods and goddesses," I said. We both glanced at the pictures of the deities taped to the wall. "The gods were always saving the devotees from wicked demons. Thakur-ma seemed so pleased with the outcome. It was as if it was happening right in front of her at that very moment. The way she told the stories, I almost believed that I could see it too—the demon's severed head rolling on the ground."

"I wish I could remember," Deepa said, sighing. "If only I'd known them when they were healthy. When they came to live with Mom and Dad, Thakur-da looked like a walking corpse, and Thakur-ma could barely talk."

As we talked, aunts and uncles (the figurative kind, not the biological) approached us and offered their condolences. We'd reply, sometimes in unison, "Thank you so much." None of the guests knew our grandmother, as she'd only recently come to the US. However, our parents had deep roots in the Indian community, so a sense of affection and obligation led to a packed hall when the ceremony finally began.

We sat in the front row next to Mom. The priest swayed and gestured as he recited sing-song prayers. He was putting on quite a show, he always did, but all his efforts couldn't drown out the noise of murmured conversations that ebbed and flowed behind us.

I saw that Deepa was crying. I could tell from how her curly hair shook when she cried. She wiped away a single tear rolling down her face with a tissue. I wondered why she was upset. She didn't really care about Thakur-ma. Every time she cried, I remembered when our cat killed her two hamsters. She must have been no older than seven at the time. I'd never seen her so upset and inconsolable. Deepa tends to let things get to her. That's her main problem.

With my sister crying beside me, I wasn't sure what to do. Finally, I put my arm around her, and she leaned on my shoulder. Her hair smelled fresh, with a citrusy, clean scent, but even that fragrance felt overwhelming. I thought about pushing her away, but decided against it. She always told me that I didn't react to things like a normal person. Now that I think about it, that's a terrible thing to say to your brother.

There was another reason I didn't push her away. This might sound a little strange, but bear with me. I rarely touch anyone, and I don't like being touched; it creeps me out. So, don't think I was experiencing some weird, perverted pleasure by having my sister rest her head on my shoulder. However, the scientific data is clear on this point: people do need occasional human contact. Touch is a powerful trigger for the release of oxytocin, a hormone that promotes a sense of well-being. I let her stay there against my shoulder because, at that moment, I could have used some extra oxytocin in my system.

As the priest droned on, I grew bored. Deepa's bony skull pressing into my shoulder started to hurt. I wished the priest would hurry up. The gods and goddesses had heard our prayers. Look at them on the wall, smiling at us, with their iridescent skin and mighty weapons. They're satisfied, so let's just finish this.

My thoughts drifted back to my farewell party, with Cohen slapping me on the back. Sanya had been there. She was so beautiful. I loved the way she wrinkled her nose when she spoke. I loved her soft, melodic voice. I loved her hair and how it bounced as she walked. But most of all, I loved those almond-shaped eyes. Yes, I'm sure it was love. That's what love is supposed to feel like. I didn't feel that same emotion again until I saw Sophie. But we'll get to that later.

Sanya and I were the only postdocs in Cohen's lab whose grant proposals weren't funded. Without research funding in academia, you're basically stuck. Sanya went back to India to be married off by her family.

I'd asked her out once for coffee, but she declined. "It's a bad idea," she told me. "We work together, you know."

Cohen betrayed me. That thought fueled my hatred. He handed over my data to Cranston at NYU to lure him into a collaboration. Then, Cranston published my data—MY data. Of course, he didn't give me any credit. Cranston wasn't an idiot like Cohen; he knew important work when he saw it. No one was discussing the use of DMTA—commonly known as the street drug Red Sky—as a psychopharmaceutical until that paper came out. Now, it's one of the most cited papers in the field. Everyone was talking about the Cranston-Cohen model of DMTA's effects on the brain. It was a revolutionary new theory about how DMTA could boost cognition and relieve depression. It even proposed a way to prevent suicides that happen in a small percentage of users. It should have, by all rights, been the Roy-Cohen model. I should've been the first author of that paper. It was MY research, and I received no credit: no authorship, acknowledgment, nothing.

Thinking about all this got me worked up, and I desperately needed a drink of water. I pushed my sister away, grabbed my plastic water bottle from under my seat, and finished it in one gulp. I squeezed the bottle and released it with a satisfying crunch. Deepa looked up at me, surprised and annoyed.

From a young age, I was painfully aware of my severe limitations. I often said the wrong things at critical moments, which brought sharp pangs of embarrassment and regret. However, I saw the path to a cure more clearly than anyone else. I knew all the major players, but they didn't understand what I understood. I had skin in the game. My entire existence was invested in this effort, and I was determined to succeed. My work would lay the foundation for DMTA to become the ultimate mood and cognitive enhancer. I knew how to do it, and no one else did. This would be the next step in human evolution, where emotions and cognitive limits wouldn't hold us back. We were so close to human perfection that I could almost taste it.

Those bastards wouldn't shut me out. They couldn't. Luckily, I hadn't shared all my results with Cohen. I reached into my pocket and grabbed the flash drive attached to my keychain. The process that would tame DMTA and turn it into a viable, profitable pharmaceutical was in my hands. DMTA, in its current form, was rightly demonized. But a new, improved version of DMTA would be the most important drug discovery ever. It would be the scientific breakthrough of the century, changing the course of human history.

I broke into a sweat again. My heart pounded wildly, like an animal trapped in a cage. I just needed McGregor to give me a shot. And for that, I needed Cohen's good graces a little longer. I'd get my revenge soon enough. My work would leave Cranston and Cohen as nothing more than footnotes in the history of DMTA. "Fuck you, Cohen. And fuck you too, Cranston," I clenched the water bottle and gave it one last fierce crunch.

"Huh?" Deepa asked. "What did you say?"

I shook my head and said nothing.

The priest stepped down from the platform, and our dad, Joyti Roy, stood up to speak.

How would I describe Dad? He was a big man, but not in an intimidating way. He wore thick glasses to compensate for his poor eyesight, which was caused by an untreated childhood eye infection.

He had a large nose and a serious expression. The word 'gravitas' comes to mind when describing his demeanor. He was always proper and formal, even with us as kids. I'd never seen him drink, gamble, raise his voice, or do anything that might be considered 'letting loose.' He was always measured and logical, almost to the point of seeming cold and emotionless. Even during heated childhood arguments where my mom, sister, and I would tear into each other, he stayed calm, detached, and diplomatic, which drove my mom crazy.

Dad stepped up to the podium and pulled out a ragged slip of paper from his pocket. He unfolded it and adjusted his glasses. "Mina, Narin, Deepa, and I welcome you on this most solemn occasion." His voice echoed through the room, silencing the dozen whispered conversations that had been taking place. "Words cannot express our profound appreciation to you for joining us and comforting us in our time of grief. Today, we mourn a woman who lived a full and blessed life. A wife, mother, grandmother—"

I whispered to Deepa, "It's almost over. We can go home soon."

"Are we having people over after this?" she asked.

I nodded. "Unfortunately."

Two

An hour later, guests arrived at our parents' house. I stood by the door to greet them, avoiding kitchen duty with my mother and sister, where the women would likely gather. Aunties touched my cheek, and uncles solemnly advised me to support my parents during this difficult time. Soon, the front hall was filled with shoes and smelled of feet and leather. By then, the pressure in my head had returned, and I looked for a space with minimal light, sound, and smell.

I moved to the living room and stood beside the round mirror with a faux gold frame. My parents bought it at a garage sale when they arrived in the US. It wasn't a good mirror; the reflected image was elongated and twisted. It reminded me of the strange visual distortions I'd been experiencing with my headaches. I wondered if these misshapen images were caused by my headaches or something else. It was odd having these—I guess you'd call them visual hallucinations. It always surprised me when it happened. The strangest part wasn't what I saw but that no one else saw it. It made me question whether I was crazy or if everyone else was. These visual distortions would appear and disappear. I had no control over them.

During my postdoctoral research with Cohen, I studied highly potent mimickers of DMTA, known as DMTA agonists. I was concerned that exposure to these agonists might be causing the increasingly strange distortions I was experiencing. DMTA addicts, called Red Skyers, hallucinate. Sometimes, they see shadows and flashes of light. Other times, they experience visual hallucinations, seeing people and animals. However, the classic Red Sky 'trip' creates a feeling that the world is alive, crimson, like a beating heart, and everything feels connected. It's a realm where forgotten memories emerge. Meaning and truth are uncovered from chaos. Many people use Red Sky—despite the risks—to find clarity and a sense of interconnectedness. They seek a place where their thoughts are clear, and unhappiness fades away.

At this point in the story, I had never used Red Sky. Even now, with Red Sky on board, I don't feel that overwhelming sense of unity with the universe. Still, I'm feeling good. I've never been so clear-headed. My thoughts seem honest and precise. I can vividly remember all the events I've described (and will describe) as if they're happening right in front of me.

Now that I've highlighted all the positives of Red Sky, I would be remiss if I didn't mention its main problem: the 'switch.' About once every 3,000 times, seemingly at random, a Red Sky hit causes the user to experience overwhelming terror. All living things around them—people, animals, plants—appear lifeless, like automatons made of plastic and metal. Red Sky users caught in a switch will tear at their skin and sever limbs to avoid transforming into these lifeless constructs. Photographs of people who died during the switch are shocking. The switch raises questions about the idea that humans want their fragile bodies to be replaced by robotic ones. It seems that humans prefer to stay human, even if it means tearing themselves apart.

Most people survive the switch. Another hit of Red Sky can turn the switch off and return the user to the classic DMTA hallucination, all fluffy clouds and unicorns. That's why Red Skyers learn to work in pairs. If one shows any signs of the switch, the other can quickly feed them another dose before the self-mutilation begins, and in most cases, that does the trick. But sometimes it doesn't. In those instances, the user and partner are found dead, killed in the most gruesome ways imaginable—eyes gouged out, skin flayed off, entrails disincorporated—you get the idea.

But here's the thing. I'd discovered something no one else in the world knew. I'd developed a compound that would prevent the switch in Red Sky users. All the details were on a flash drive in my pocket. I was sitting on a goldmine.

Here's the plan: combine the drug I developed with a low dose of DMTA into a single pill. Then, people can take it and enjoy all the benefits of Red Sky without worrying about self-mutilation.

I was close to developing a combination drug that would allow anyone to be intelligent, creative, and happy, or even get high, whenever they wanted. DMTA would still be highly addictive. However, anecdotal evidence suggested that microdosing Red Sky greatly reduced the risk of addiction. By microdosing with Red Sky alongside the drug I developed, one could become a DaVinci-level genius without worrying about harming oneself.

We needed to find the correct dose of Red Sky to prevent addiction. That should be straightforward. Once my combination drug was ready, everyone would want to try it. Why wouldn't they? It'd be a huge success and change the world.

There were "true believers" who turned using Red Sky into a quasi-religion. You saw them everywhere—walking around, mumbling, and staring at the sky. They're happily addicted and wouldn't want it any other way.

But imagine, with just a single small combination pill, you could experience mental clarity like never before and prevent DMTA-induced psychosis. That's what the medical field calls the switch, 'DMTA-induced psychosis.'

My heart pounded at the thought of how wealthy and famous I would become once I developed my wonder drug. I needed lab space, a few months of dedicated time, and funding to start a pharmaceutical company. That isn't too much to ask for.

I looked again at my distorted reflection in the mirror. Maybe I was going crazy, which was a genuine concern at that moment. I was seeing some strange stuff even without the migraines. But then I reassured myself: I've done my research. Crazy people, like those with schizophrenia, have auditory hallucinations—they hear things that aren't there. If the hallucinations stayed purely visual, I wouldn't have anything to worry about. I just needed to hold it together a little longer. That's all. Hold it together. I was so damn close.

Despite the distortion, one thing was clear—I looked like a total mess. I ran my fingers through my hair and straightened my crooked tie. As I did, I heard my father calling me from across the room. I walked over to him.

My dad said, "Professor Goldblum asked that I introduce you to him."

I nodded at the white-haired, older man who was holding a saucer in one hand and a teacup in the other. Professor Goldblum was nearly a foot shorter than me. He wore a brown sweater vest and green corduroy pants, tightly fastened high above his waist. He straightened his stooped posture, leaned in, and studied my face through his narrow-rimmed glasses.

"My, you have grown. When I saw you last, you were this tall," Goldblum said, lowering the saucer to his waist. "You don't remember me, I'm sure."

I nodded.

A smile spread across Goldblum's face as he lifted the teacup. "Have you tried this fine tea your mother serves in the kitchen?" The tea gently quivered in Professor Goldblum's trembling cup. "Almond and cinnamon, it's your father's favorite." My dad nodded in agreement. The professor took a sip, and steam fogged up his glasses. "Two ordinary flavors blended to create an extraordinary taste. As with tea, so it is with friendship. The sum is greater than the parts."

"Indeed," my dad said. "The professor and I were friends even before Deepa was born."

The professor moved closer and spoke to me in a conspiratorial tone. "But you, Narin, should realize that I remain unconvinced that your father's tastes are better than mine. Surely, this is just stubbornness on my part. I am, in fact, quite stubborn by nature. But how should I prove my superior taste?" The professor looked at Joyti and continued, "I've given this some thought. Perhaps more than it deserves. And I have reached this conclusion. You, young man, are uniquely qualified to settle this once and for all." He looked up at me. "And so, I have a small request. When you visit St. Louis next week, please take a moment to share a cup of tea with me. I would really appreciate the company and look forward to hearing your judgment on what may seem trivial to you."

I smiled and nodded politely. I didn't mention that I had no plans to visit my dad's old friend when I went for my job interview.

As the night grew late, the crowd started to thin out. Mom pulled Deepa and me aside and told us we would stay overnight to help sort through our grandparents' belongings in the morning. She explained that she had been quite tolerant of her in-laws, allowing their junk to pile up in my room. But now that they were both gone, everything had to go. We reluctantly agreed to stay, even though we didn't really have a choice. Mom wasn't taking 'no' for an answer. She's the most volatile person in our small family, and we didn't want to upset her.

Mom was about Deepa's height, and she had become stout and gray with age. Her mouth was always in a frown. Although she had many friends in the expat Indian community, she constantly complained in private about what so-and-so said or did. Deepa and I had always felt she thrived on the drama of petty disputes among friends. Why else would she get involved in every little conflict that arose?

"I haven't slept in my bed for years," I told Deepa.

"I slept on mine last Christmas," Deepa replied. "I'd forgotten how uncomfortable the mattress was until I had to sleep on it again. I can't believe I survived my childhood."

"At least no one has died in your bed recently," I said.

Deepa went to the kitchen to help with the cleanup, while I stayed in the living room, saying goodbye to those who were leaving. In the gold-rimmed mirror across the room, I saw the dark reflection of two men in gray suits. They huddled together and occasionally looked in my direction. They didn't resemble the other mourners; they seemed more like they were heading to a business meeting than a funeral. I wasn't interested in learning more about them. I winced as they moved toward me. With no clear escape route, I folded my arms and pressed against the wall as they approached.

Ian was the first to arrive. He was tall and thin, with white-blond hair that lifted and curled like a bird's plume. He extended his hand. "Hello, I'm Ian, Ian Blair. I'm an acquaintance of your father." I shook his hand, and he held it tightly, almost painfully. He flashed a quick smile. Standing with his shoulders back, chest out, and chin raised, he said, "I'm the Chief Counsel for Harvester Pharmaceuticals." Maru Chandra here is our Vice President of Drug Development." His companion was a dark-skinned, thin, nerdy-looking Indian guy. He didn't look at me, didn't extend his hand or smile. He seemed to be brooding, as if I'd somehow offended him.

As far as Ian was concerned, I wasn't sure what to think of him. "How do you know my father?" I asked.

"He was my faculty advisor when I was an undergrad before I went to law school," Ian said. "You don't remember, but I once had dinner at your house."

"You looked familiar." I tried to smile, but my lips only tightened.

Ian continued, "I heard about your grandmother's passing and wanted to come and pay my respects." He motioned to Maru. "And as luck would have it, Maru and I have an ulterior motive for coming today. We wanted to meet you."

"Meet me?" I stammered. "What do you want with me?"

"Your paper in *Nature* caused quite a stir at Harvester," Ian lowered his voice. "We have a group of scientists very interested in exploring the therapeutic potential of DMTA. That is your area of expertise, isn't it?"

"I do have an interest in that." I rubbed the back of my neck. Something about Ian made me want to pull away and run as fast as I could. For one, he stood way too close, invading my personal space. But it wasn't just that. I felt as if he were trying to draw me in, softening his center to pull me close, surround me, crush me, and leave me an empty shell. I'm not sure I'm explaining this well. He filled me with a vague sense of dread. Another way to put it is that I felt creeped out by him, but I wasn't sure why.

Ian nodded, moved closer, and lowered his voice. "The thorny legal issues surrounding DMTA make it challenging to do the work that we do. But you'll find that Harvester is an extraordinary place. We can get things done in a way others can't."

"I know," I said. "I saw that on the news."

Maru bristled.

Ian threw his head back and laughed. He gripped my shoulder. The last thing I wanted was for him to touch me, so I pulled away from his grasp. "What we do at Harvester would amaze someone like you," Ian said. "I overheard that you're coming to St. Louis next week. We'd love to show you around our newly renovated international headquarters and introduce you to some of our top scientists. Perhaps we can even schedule a meeting with our CEO, Sanjay Chandra, who is also Maru's father. He was one of the founders of Harvester back in the day." Our eyes shifted to Maru, who looked away. Ian continued, "I'm sure someone with your expertise would be interested in what happens behind the scenes at a place like Harvester Pharmaceuticals. It's not such a bad place to work." He released his grip and handed me his business card. "My cellphone number is on the back. Give me a call when you get into town."

"Thanks for the invitation." I needed to get away and be alone. "I'll see what I can do," I replied and hurried away.

Three

I retreated to my childhood bedroom and closed the door behind me. But this wasn't the refuge I was seeking. The floor was covered with piles of ragged clothes that smelled of incense and old age. I had always kept my room spotless, but the sight and smell of the mess made me feel physically ill. I couldn't understand how my mother had let my room become so filthy. She knows that dirt attracts bugs, and she hates bugs as much as I do.

When I was nine, my dad called me 'fastidious.' I didn't understand what that meant back then, but I do now. Yes, I admit, I'm fastidious in both ways. I am very concerned about cleanliness and pay close attention to every detail. That's why I want to write down my story while I still have Red Sky on board and can easily visualize everything that happened. This might be my only chance to get everything right.

Knick-knacks cluttered every corner. There were pictures of saints and deities, stacks of file folders filled with yellowed papers, and a bulky, dirty comforter shedding cotton fluff through holes. My grandmother had made a makeshift shrine on the dresser with the wooden icons she brought from India. I opened the window to air out the room.

The cold April breeze swept away most of the smell. I listened to the rustling leaves of the big maple outside my bedroom window and searched for a blanket, but I couldn't find one that didn't smell like urine and cigarettes. I decided I'd rather be cold than suffer a horrible migraine. I sat on my bed, feeling sad as I looked from one trash pile to the next.

I started shivering, so I began organizing the piles to stay active, move around, and stay warm. But I soon gave up. It all just needed to be tossed. There were so many pictures of gods and goddesses. Do we really need so many deities in our lives? I took a deep breath and shivered again. I tried to think warm thoughts, like my grandmother being cremated a few days earlier. It must have been warm in there. The smoke rising from the crematorium carried her to the heavens to be with her deities. Good thing, too. They were her true family, the divine beings she worshipped with all her heart. She'll be much happier with them than with us mere mortals.

I sat on the bed again, and the worn-out mattress springs squeaked beneath me. I remember those sounds well. As a boy, I had to be careful about what I did in bed. I did it quietly so the noise wouldn't expose me. I worried my parents could hear me, and the slightest squeak would reveal my shame.

There was a foul smell coming from the bed. After my grandmother's death, my mom stripped it and made it up for me with "fresh" sheets. The pressure in my head increased until it felt like a pounding headache was about to set in. Spending a night on those musty sheets, which had probably been sitting in the linen closet since I went to college, was sure to bring on a full-blown migraine.

I stripped off the sheets and pillowcases and threw them against the far wall. Then, I went to the bathroom and gathered a stack of clean towels. With these, I covered the mattress from top to bottom. The mattress itself was stained and didn't smell great either. I wondered whether that was my fault or my grandparents'. Despite their bleachy smell, the towels were better than the stale sheets and the disgusting mattress.

I grabbed a black button-up sweater from a nearby bag and pressed it against my chest. My reason was simple: I was freezing, but I couldn't close the window or I'd suffocate. Sadly, the sweater was beyond repair. Frayed strings hung from the sleeves, and the stale cigarette smell instantly brought back memories of my grandfather. I threw it against the far wall, where it landed on top of the pile of sheets. I needed something to keep me warm. I'd worn a suit to the funeral and really couldn't sleep in that.

I looked through a bag, found a wooden box, and turned on my bedside lamp to get a better look. The lid showed an elephant carrying a bearded nobleman riding in a fancy carriage on its back. The box smelled like sandalwood, a nice scent that didn't give me a headache. I opened the lid and took out a bundle wrapped in newsprint. I tugged at the old string holding the bundle together. When I touched it, the string fell apart. I peeled away the layers of paper, just like peeling an onion. To my surprise, inside was a knife.

I carefully lifted the knife by its carved wooden handle and examined it under the light. My stomach clenched, and a chill ran down my spine. It was beautiful. The intricately carved handle tapered into a leaf-shaped blade that glowed deep red, like dying embers. I saw my distorted reflection in the blade. It looked like me, but somehow different. I had an unsettling feeling that I was looking at a better, more confident version of myself. I was seeing a Narin Roy that I wished I could be. This Narin would have stood up to Cohen and demanded his due instead of smiling meekly like a damn idiot.

I was familiar with the story of the knife. My grandfather told me this scandalous tale. According to him, this very knife had been used by a wealthy Indian businessman to kill his daughter and a visiting American student when he suspected they were having an affair. He shared this story one night, shortly after they arrived from India, while we sat on the porch. He recounted the gruesome details between puffs of smoke from his cigarette and fits of coughing. After he finished, he gave me a long speech about how the murderer was "absolutely right" to do what he did—killing his daughter and her lover—something about honor, or some other such nonsense.

Since that night, I often thought about his story. I tried to picture the poor daughter and her—maybe her lover? What made her betray her family and fall for this American? He must have seemed exotic to her. His experiences would have been so different from hers. He would have seemed so worldly compared to the other boys she knew. Was she rebelling against a system of oppression, or did she lose control, with one thing leading to another? Did she regret the affair, that is, before her father found out? Maybe she was trying to end things with him when her father discovered them together. I tried to imagine the horror she must have felt when she saw her father raise the knife against her. It must have seemed unreal, like a terrible nightmare.

I fell into a strange reverie, gripping the knife handle and wielding the weapon menacingly. What drives a man to do the unthinkable? Anger, hatred, jealousy, pride? What happens when these intense feelings fade? How could you live with yourself afterward? I brought the knife closer and felt the cold blade against my thin T-shirt. What would it be like to slide it between my ribs? I could do it effortlessly. The thought both terrified and exhilarated me at the same time. I believed I could do it. No one could stop me. The signal would travel from my brain, down my spine, and activate my motor neurons. The muscle fibers would obey my commands, with actin and myosin sliding past each other, causing my biceps to contract. What was stopping me? It would be over in a moment. I had worked tirelessly over the past few years, only to suffer through countless hours with little to show for it. But something kept me from ending my life. I was still in the game. I had a major discovery in my pocket. Still, if things went wrong, I had this knife as a backup.

A knock on the door interrupted my thoughts. "You still up?" Deepa asked.

"Just a second." I hurried to put the knife away, kicking the knife box under my bed with my heel. "Okay, come on in."

Deepa stepped inside and looked around. "Shit," she said. "What a disaster! What are we going to do with all this junk?"

"Ten large garbage bags and a dumpster fire," I said.

She went to close the window, but I told her I needed to air out the room longer. She shrugged as if to say, 'If you want to freeze, that's your own business.'

She grimaced as she looked from one pile to another. "I thought we were going to sort through all this stuff tomorrow," she said.

"I already tried. There's nothing worth keeping. We can take another look tomorrow. I'm beat." I was hoping she'd take the hint and leave. I was eager to take the knife out again and examine it more closely. I still felt the imprint of the carved handle on my hand.

Instead, she pushed my desk chair toward me and sat down. "You know why I want to talk to you, right?" she asked, glaring at me. I wondered what I'd done. "Mom and Dad aren't speaking to each other," she said.

"I wasn't aware of that," I replied flatly, hoping to cut the conversation short. But this only seemed to infuriate her. "Why?" I asked lamely.

"Why the hell do you think, Narin?" she yelled.

I hated it when she yelled at me, even more than when Mom yelled. I looked down at my hands and tried to seem pitiful. "Well, it's Dad's fault," I said. "He should never have brought them here."

"Does it matter whose fault it is?" She was still yelling. "Thakur-da was dying of cancer. Dad was misguided, but I don't blame him for wanting to help. I know how Mom feels about it, but I would've done the same thing."

"When Mom's mother was dying of cancer, they didn't try to get any specialized treatment for her," I countered.

"They were poor back then," she shot back.

"They're not rich now," I muttered. The louder she got, the quieter I became. I didn't want an argument, but Deepa was being unreasonable. She always sided with Dad. Mom had every reason to be furious. Dad had spent a large part of their savings to bring his parents to the US without telling her until they were nearly at her doorstep.

We sat in silence. I noticed Deepa's face was flushed, which was a bad sign. I kept staring at my hands as the imprint from the knife handle quickly faded.

"This is what we're going to do," she said firmly. "Tomorrow, I'll talk to Mom, and you'll talk to Dad. You ask Dad to apologize for not considering her feelings. I'll speak with Mom and ask her to forgive him.

I think it's important for them to reconcile before they go to India, and you leave for your job interview. It's dangerous to let things fester. The sooner they reconcile, the sooner we can get back to the happy family we were before our grandparents arrived."

We sat in silence once more.

"You know, there's a history behind all this," Deepa said. "Thakur-da refused to bless their marriage. Mom never forgave him for that. Didn't you notice how rude she was to him after they arrived last fall?"

I hadn't realized that either, so I stayed quiet.

"For a smart guy, you're a total idiot sometimes," she spat.

Tears welled up in her eyes. "I'm sorry," she said. "I didn't mean that. I'm not mad at you. I hate to see them both so upset. I just wish that you'd—I don't know—notice these things so you could help me get everything sorted out."

I looked up at her and said, "I think the proper term is 'idiot savant.'"

She laughed, wiped her eyes, jumped up, and hugged me. "Okay, idiot savant, make sure you talk to Dad tomorrow."

She pulled back. "Oh my God, you're bleeding!"

I looked down and noticed a large blood stain on my T-shirt. I lifted the shirt, pretending to check where the blood came from but trying to hide the hole I'd made with the knife. Sure enough, there was a cut with dried blood right over my heart.

"I must have cut myself while I was going through their stuff," I muttered, waving a careless hand toward my grandparents' junk. "It's just a scratch." I tried to laugh. Even without looking at her, I could tell she knew I was lying.

"I'll bring you a bandage."

She soon returned, knelt on the ground, and wiped the wound with alcohol, which stung like hell. She ripped open the bandage and placed it over the cut. I wondered what she was thinking. I must have looked so small, pathetic, and clueless.

"We should hang out more often," she said softly. "I barely get to see my big brother anymore. You know I'm here for you if you need me. Right?"

I nodded but didn't dare look at her. I love my sister; she has such a good heart. I knew she wanted to ask me if I was still taking my meds. I was glad she didn't because I didn't want to lie. I wasn't sure what kept her from asking me. I had been off my medications before, with disastrous results when I was a teen and again in college. She was stressed about starting medical school and managing all the family responsibilities. She didn't want to deal with a brother who was falling apart on top of everything. Admitting to her that I was slipping into another bout of severe depression would mean she'd have to make sure I saw a psychiatrist and a counselor. Neither of us could handle those responsibilities right then.

After she left, I considered taking out the knife again to examine the handle more closely. Instead, I turned off the light and lay on my back. That knife had already caused me enough trouble.

The wind was picking up, and the rustling of the maple leaves sounded urgent. I had to choose between freezing or dealing with a migraine. I decided it was better to freeze. I grabbed a towel, wrapped it around my shoulders, and stared at the glow-in-the-dark stars on the ceiling. It had been such a long day.

I was just about to fall asleep when I heard it. "Narin," someone called out from the darkness. "Your life is in grave danger. But rest assured, dear boy, that I am here to help you. I will prepare you for the coming conflict. You shall be ready when the fateful day arrives."

I sat up and turned on the light. No one was in the room. Had the voice come from outside? No, it sounded like it was inside. It felt as if it was coming from inside my head.

Are you fucking kidding me? Now I'm hearing voices. Should I panic? With that thought, I lay back down and fell into a deep, dreamless sleep.

Four

The next morning, I woke up feeling as cold and stiff as a corpse. Most of the bath towels were on the floor, and I still wore the suit I had on the day before. I wandered into the kitchen, feeling bleary-eyed, and made some toast and tea. As I finished my breakfast, Deepa found me.

"Hey, big brother," she said, a little too brightly for my taste, "how about we get started on our huge cleaning project?" I glumly followed her back to my room. For several miserable hours, we sorted through piles of what could only be called junk. Although not usually religious, Mom felt it would be wrong to throw away religious icons. We packed up the wooden and metal figures of gods and goddesses and collected the colorful pictures, which were then placed in my old sock drawer, where they would probably stay forever. The rest of my grandparents' belongings were bagged up and hauled unceremoniously to the garage for disposal. Deepa kept a few old black-and-white photos and Polaroids. She didn't know any of those smiling relatives, but she thought it would be wrong to discard them. She and my mom are alike in many ways, including their sentimentality.

After finishing those tasks, Deepa and I went to the kitchen to warm up a frozen pizza. We were both surprised to see that there wasn't any Indian food in the fridge. When we were kids, there was always something: rice, chickpeas, lentils, chicken curry, or at least some leftover naans. But the fridge was nearly empty, so we ate whatever we could find.

"I went to this med school meet-and-greet last weekend," Deepa said enthusiastically as I pulled the pepperoni slices off my pizza and put them on her plate. "And then a group of us went to a place called the Cine afterward and watched a movie that came out last year, called *Parineeta*. Oh my gosh, it was one of the best movies I've ever seen!"

"Uh-huh," I replied noncommittally. You might think I'm crazy for giving away my pepperoni, but I don't like anything greasy. It makes my mouth feel weird.

"The music, the choreography, and, my goodness, the acting, it was incredible!" she gushed.

"Sounds great."

She scoffed. "I know you don't like movies, but I think you might like this one."

"What's the point?" I replied. "It's just a made-up story." I wasn't a fan of movies, especially musicals. I can suspend disbelief when watching an average film, but when they burst into song and dance, that's too much for me.

She ate the pepperoni slices, wiping her fingers after each one. "During the movie, I sat beside this really cool guy," she said nonchalantly.

Finally, she said something interesting. I glanced over and saw her blush. "Tell me more about that," I said. I didn't care about her romantic relationships. As I mentioned, she had only marginally more success with the opposite sex than I did, even though she was attractive and acted like a normal person. Still, I was always looking for something to needle her about, and a new boyfriend seemed like just the thing.

"Well, he likes many of the same movies I do and knows all that old-school stuff. He's seen all the Amitabh Bachchan movies."

"Hmm, that sounds promising."

"When he found out I was Bengali, he began quizzing me about Satyajit Ray's movies. He rattled off three or four that he said were brilliant and that I needed to see." She pressed her lips together and appeared deep in thought.

"He sounds interesting." I watched for a reaction.

"His father is a cardiologist in Los Angeles," she said.

"Wow," I said. "He sounds like a real catch."

"Shut up." She punched my arm. "He just seemed cool, that's all."

"Did he ask you out?"

This got the reaction I wanted. "No," she said, wincing. "But there's another one of these meetups next week, and he said he'd be there."

She grabbed a handful of napkins and wiped her hands. "So, when are you going to talk to Dad?"

"As soon as I finish eating," I grumbled. She had used nearly half the napkins to wipe her fingers. What a waste.

I washed up and went to my dad's study. He was sitting behind his oak desk. I noticed the dark circles under his puffy eyes. His head hung low, and his thick glasses slipped down the bridge of his nose. Soft Bengali music played on a cassette player on his desk. I've heard these songs a million times, mainly in the back seat of the car on long road trips, so the music made me feel a little carsick. He held a small wooden monkey. As he examined it, a one-sided smile appeared on his face.

"Hi, Baba. Did you get the tickets to India?" I asked.

He placed the wooden statue on his desk. It usually sat on the shelf behind him. When I was younger, I often took it down to examine it. The figure was roughly carved, but its human-like face always fascinated me. As I stared at it, the face seemed to change expression—a smile appeared on the little face. I blinked and shook my head to clear the image. When I looked again, the figurine had returned to its frown. But this wasn't the time to get distracted. I needed to focus on this conversation with Dad, or Deepa would kill me.

"The tickets are quite expensive, even for coach," he said. "Unfortunately, there are business matters that require attention. Your grandparents left quite a mess."

Tell me about it.

Dad continued, "Cousins are coming out of the woodwork wanting me to sign away my rights to the business and house, which I will gladly do."

"You can't do that online?"

"There's also the matter of the ashes," he said. "Besides, your mother wishes to visit her father."

I kept glancing at the monkey statue, tempted to pick it up. It had smiled at me, after all. But what if it decided to do a jig in the palm of my hand? Considering how things were going, that seemed more likely to happen than not.

He asked, "What kind of questions do you expect at your job interview?"

The thought of the interview made me panic, but I kept my voice steady. "I've reviewed all my papers and abstracts," I said. "I've memorized a few key points Professor McGregor will probably ask me about. It should be a friendly interview. He'll ask about my postdoc work here in Chicago and what I expect to do in the future. He and my current advisor, Dr. Cohen, are good friends."

My mouth felt dry, and I cleared my throat. "Due to the Red Sky epidemic, the NIH plans to issue an RFA in my area of expertise. That will make it easier to secure funding the next time I apply for a grant."

He adjusted his glasses. "RFA? What is that?"

"Request for Applications," I said. "That means they will call for new research in my field of interest, specifically the chemical compound DMTA."

"Why is DMTA called Red Sky?" Dad asked.

Oh gosh, he's asking the same old questions. Does he not remember anything I told him? Is he getting Alzheimer's disease?

"One of its common effects is that it makes the sky look bright red," I replied.

Dad nodded solemnly. "I recently read an article about Red Sky." His eyelids drooped, and his eyes lost focus for a moment. I'd never seen him look so exhausted, so defeated. "It's quite tragic. So many communities have been devastated, and so many lives have been lost."

"It's becoming a serious problem here, too," I said. "If you go downtown, you'll see the addicts leaning against the buildings, just staring.

But most users don't have an addiction. Most people use it when they need a boost."

"But why do they take their own lives?"

'Why indeed?' We had this exact conversation about a year ago, but I answered his questions as if we'd never discussed it.

"It has to do with how the drug affects the brain's chemistry," I said. "The short answer is that no one knows. That's one of the central questions in my grant proposal. Dr. Cohen and I have developed a model to explain what's happening in the brain. It's related to the effect of DMTA on two newly discovered brain cell receptors. I want to validate some of his ideas and try to put them on more solid ground."

He frowned. "Is this chemical DMTA the same compound secreted by poisonous toads?"

"It's similar," I replied. "But DMTA is much more potent. It's about a hundred times stronger than the hallucinogen secreted by Sonoran Desert toads. In pill form, DMTA is easily produced and distributed. That's part of why it's become so popular, not to mention that it makes people euphoric, even in small doses. In that way, it's very different from other hallucinogens. That's what makes it so addictive."

He shook his head. "It is sad that such things have become commonplace today."

I felt compelled to respond to this. "It is sad that so many people are addicted and dying. But it's also interesting that this simple compound can profoundly influence people's perceptions and behavior. They've identified specific receptors in the brain that bind to DMTA. These receptors must exist for a reason. They wouldn't have evolved if they hadn't helped us survive." I began absentmindedly pulling at my hair, a nervous habit. "No one has yet determined what these receptors do. Why does the brain even have receptors that bind to DMTA? The answer to that question could lead to breakthroughs in understanding how the brain processes information. That's why this work is so important. It could be the key to understanding schizophrenia, autism, depression, and so many other terrible conditions. DMTA, as a mood and cognitive enhancer, is incredible. There is still a lot to learn about what it does in the brain to produce these positive effects."

"I guess everything has a potential benefit." Dad took a deep breath. "So, do you think you'll get the position in St. Louis?"

"Dr. Cohen thinks I will," I said, feeling the panic rising again.

"What if the interview doesn't go well? Is there any way you could find a job here so you can stay in Chicago? I'm not asking for myself but for your mother and sister. Having you nearby during this difficult time would be helpful."

I shrugged. "Believe me, I've looked, but I haven't found anything worth pursuing." That's when I remembered the conversation with Ian and Maru. "Oh, but I did get another lead yesterday. One of your former students, Ian, approached me. He's with Harvester Pharmaceuticals. It sounds like they might be interested in hiring me. I've never considered working in industry before, but I guess it wouldn't be the worst thing in the world. Maybe I could do that for a while until something better comes along. For some reason, they're interested in DMTA."

"Ian mentioned the same to me." His eyebrows knitted together, and his forehead wrinkled. "If the job involves working with Ian, you may want to consider carefully before you agree."

"Why is that?"

"Ian's an interesting fellow," he said. "When he was a student here, I was his faculty advisor and got to know his unique circumstances." He paused, and I waited for more.

Dad bit his lower lip, swallowed, and continued, "Unfortunately, he suffered a traumatic childhood and missed a fair amount of school to receive counseling. That's not particularly unusual. Many students become emotionally distraught in college and find it difficult to deal with past trauma in a new environment. I wouldn't judge a person's character based solely on this. But there were also accusations against him during the short time I knew him that suggested a very disturbed personality."

I moved my chair closer. "What kind of accusations?"

"It's best not to discuss such matters," he said. "Maybe I've already said more than I should. After all, I didn't investigate the truth of the accusations myself. That wasn't my role. I completed the paperwork and provided guidance when requested. Still, the charges against him were serious enough to justify his expulsion. He was only reinstated after the threat of a lawsuit."

"When did this happen?"

"About seven years ago." He took off his glasses and wiped them clean. "He was just a young man back then, and his temperament may have changed. By all appearances, it seems he has done well for himself. He's a successful lawyer now. Perhaps he has learned to subdue his demons or channel them differently."

I thought about Ian, with his perfect blond hair and striking blue eyes.

Deepa ducked her head in. "Baba, Ma's crying in the kitchen! You need to go talk to her right now."

Dad sighed, pushed back from the desk, and turned off the cassette player. Then he slowly walked up the stairs and into the kitchen. We followed him and waited outside the door.

"Did you talk to Mom?" I asked.

"Yeah," she whispered.

"Strong work, Sis. You did great," I said more harshly than I meant to.

I thought she would yell at me for not caring about our family and only thinking about myself. But she didn't. She stood beside me, struggling to hold herself together. She always lets things get to her. I felt terrible for her. I hoped she could make it through medical school and become a doctor, caring so deeply about the suffering of others. It's her biggest weakness. Whatever happens, she'll need to become less sensitive, or she's likely to drive herself crazy.

Don't get me wrong. I'm not a misanthrope (I hope I'm using that word correctly). I wouldn't have pursued DMTA research if I didn't care about humanity's future. Still, my love for humans is purely intellectual, not emotional. On a personal level, I can't say I've ever loved anyone except my family, Sanya, Ian (as a friend), and Sophie (but we'll get to that later). As you might guess by now, I've always avoided getting close to people.

We stood there, listening, but we only heard Mom.

Mom yelled, "So, I have no voice in this matter? So much of our money is gone now, wasted on your worthless father!"

After a moment of silence, she continued, "Of course, you had a choice! Don't say that you didn't have a choice."

I whispered, "I don't think we should be listening to this," but we didn't move. If it had just been me, I would have left. But the truth was that I was afraid to leave Deepa alone; she seemed so vulnerable.

"Don't you dare defend that man after he tortured you, slandered my father, and then refused to bless our wedding!" she yelled.

A long silence followed.

"And what did that get you? Tell me that. He was so pigheaded that he refused all the treatments you paid for with our hard-earned savings!"

Another silence followed.

Mom shouted, "Don't you see what's happened to our children? Don't you see how devastated Deepa is when she needs to focus on her studies? Don't you see how gloomy Narin has become? They are the ones who need our help, not your filthy father."

"She does have a point there," I said. I know I shouldn't make light of a terrible situation, but that's how I cope. "Well, at least they're talking now. That's something, huh?" I know that was another stupid thing to say. I always say the wrong thing and make things worse. I wanted to put my arm around Deepa and tell her everything would be okay. But I didn't want to—I couldn't—I just couldn't lie to her like that.

Five

It's July 2006 as I write this, and I'm sitting in a police interrogation room in St. Louis. Everything I've described so far happened in April. This journey started just three months ago, but it feels like a lifetime.

The U.S. had been—and still is—at war in Iraq and Afghanistan, and according to news reports, things weren't going well. Our enemies were gaining the upper hand, and American soldiers were dying every day. They were being blown up by improvised explosive devices and ambushed in surprise attacks. They fought street by street in cities that are well known to our adversaries. Public opinion was turning against the war, but the fighting continued.

The government seemed increasingly desperate to gather actionable intelligence (that's how they described it) against this elusive enemy that hid in plain sight, disguised as civilians, and only revealed themselves when they pressed a button, indiscriminately killing soldiers and civilians alike. Graphic images of men, women, and children torn apart flashed across our TV screens every night. All of this felt worlds away and was not my problem. Like most Americans, I went about my daily life as if everything were normal. I had no idea I would soon be caught up in the chaos on the other side of the globe.

May was approaching, but the days remained cold and the nights colder still. I returned to my quiet, tidy apartment to prepare for my trip to St. Louis. The weekend went by, and no more voices warned me of my grim fate. I still experienced strange visual distortions, sometimes accompanied by migraines, but often they occurred without headache. Sometimes, it was just a flash of light. Other times, I felt like everything around me was unreal. However, these sensations only lasted a few seconds or at most a minute, fading as quickly as they appeared. As strange as it sounds, I kind of wished I was actually hallucinating—like a group of beautiful women in my room staring at me with admiration and desire. But no such luck.

I only had one good suit, which I had just worn to my grandmother's memorial service. So, I had it cleaned, even though it probably didn't need it. My apartment lease was about to end, but I had already planned to move, so I wasn't worried. My parents left for India the day before my flight to St. Louis. That weekend, I expected at least one call from Deepa to check on me. But nothing. I guess she had gotten too busy to worry about my well-being.

I reached out to my friend Venu, a hospital pharmacist in St. Louis. We hadn't spoken in a while. He was one of my few close friends in college, but like many college buddies, we had grown apart. He had gotten married and had a child. From the tone of his voice, he didn't seem eager to host me, but he finally agreed without complaint. I wasn't expecting much from him except a place to sleep before my interview.

I flew out Monday morning. My flight went smoothly, and the landing was gentle. Venu picked me up at the airport, driving a practical silver minivan with a booster seat in the back. He now seemed to embody the typical 'family man'. As he drove up, I wondered what we would talk about on the way to his house. We no longer had much in common.

He had a receding hairline in college, and now he was almost completely bald. He also had a potbelly, which was common among some Indian Americans. Deepa was so worried about this that she was always doing planks, Pilates, and yoga.

I got into the passenger seat. "It's chilly for spring," I said. "It's as cold here as it is in Chicago."

"It's been raining," Venu replied. "How have you been?"

Boy, that'd be a long story.

"Just ready for the interview," I mumbled. "I can't wait for it to be over."

"Have you ever been to St. Louis?"

"Once, for a couple of days, a long time ago," I said. I remembered seeing the Gateway Arch as a kid, so I must have visited St. Louis at some point.

"It's not bad," Venu said. "It's not Chicago, but the traffic is much better, and the housing is more affordable. It grows on you."

I asked, "How's your job going?"

He shrugged. "It's a paycheck. That's all."

We rode along in silence for some time.

"I'm sorry about your grandparents," Venu said. I was surprised he knew about that. I hadn't mentioned it to him. "Were you close to them?" he asked.

"Not really," I replied. "My grandfather was cool. He taught me some Bengali obscenities, which drove my mom crazy. Don't get me wrong, he was a terrible human being with racist and sexist beliefs, and he was a horrible father to my dad. But I liked him all the same. He could tell a dirty joke as well as anyone."

"What about your grandmother?" he asked.

"She just died. She died in her sleep a week ago. I guess we're all still in shock. With my grandfather, we knew he would die soon from his lung cancer, so it wasn't a big deal." I explained that my dad had brought them from India so my grandfather could get specialized treatment.

"How long were they with you?"

"They came last fall. He died in January, and she died last week."

Venu frowned. "You should have rescheduled the interview with everything going on."

I couldn't help but sigh. "The training grant expired a couple of months ago. My R01 received a good score from the NIH but wasn't funded. My advisor has been covering my salary, but I need to find a job that offers protected time to gather more data and revise my proposal. I need to resubmit my grant in the next cycle, or I'll never get funded."

"I'm so glad I don't have to deal with all that crap." Venu shook his head. "I never understood why you went down that route. It's much easier being a pharmacist and earning a steady paycheck. You don't have to worry about chasing after grants or losing your job."

"Sometimes I wonder." I thought about how the wooden monkey had changed its expression and the other strange visual distortions. I squeezed my eyes shut to push these thoughts away. "I've worked incredibly hard over the last few years," I said. "God only knows what I've done to my health after being exposed to so many DMTA-like compounds."

"You're still young," Venu said. "You can do anything you want."

I ran my hands through my hair and felt my heart pounding. I used to pull at my hair when I got agitated, but I'd promised myself and my former therapist that I would resist the urge. Doing it too often can damage the hair follicles, and eventually, your hair will fall out.

"I wish I could," I said. "I feel like I've been at this so long that I can't imagine doing anything else. It's all I think about. You could almost say I'm obsessed." I let out a small laugh. Venu knew all about my mental health issues. He had seen me at my worst when I was hospitalized in college for suicidal thoughts. He'd witnessed my unraveling when the stress of getting good grades sent me into a downward spiral. Yeah, I'm sure Venu still saw me that way, in my blue paper clothes in the psych ward, wild-eyed and crazy. Who could blame him for having reservations about my staying with him and his precious family?

I was fidgeting now. "You know, the World Health Organization just named Red Sky addiction as the number one public health problem in the world. Think about how much better this world would be if we could control the effects of Red Sky so people don't get addicted or go into the switch."

"Why?" Venu shrugged. "Why waste time and money to save a bunch of junkies?"

I didn't think I could get any more agitated, but his last statement proved me wrong. I grabbed a fistful of hair and tugged at it. "DMTA isn't like any other street drug," I shouted. I didn't mean to yell, but I did. "It's not like meth or heroin or cocaine. Red Skyers aren't irrational or uneducated. Some people take it too far, but some of the world's most brilliant and talented individuals routinely microdose with Red Sky. They say it opens their minds and lets them access the deepest parts of their psyche. Artists, actors, writers, and CEOs regularly use Red Sky to boost their creative energy."

"So, you're saying Red Sky is some kind of wonder drug?" He rolled his eyes.

"It could be a wonder drug," I said. "That's exactly my point. But the key is understanding what's happening in the brain. What exactly is DMTA doing? If we understand that, we might be able to revolutionize our approach to mental illness or even everyday problems of living. Why do some people seem to have better memories than others? Why are some people so creative? Why are some people anxious or depressed?" He looked at me, probably to see how far gone I was, but I continued, "Ultimately, it all comes down to neurochemistry. Consider all the good we could do if we could freely explore everything hidden in our minds beyond our limited awareness. Who wouldn't want a pill that boosts creativity and helps you tap into your mind's hidden depths? This could be the key to solving all the world's problems."

"The same things have been said about LSD and marijuana for years," he said. Before I could respond, he continued, "Red Sky doesn't enhance your awareness. It makes you hallucinate. It doesn't improve your insight. It destroys it. It's just like all other street drugs. Trust me, I've seen them when they come to the hospital. It's as if they've completely lost touch with reality. You can't reason with a Red Skyer. To them, you're like a shadow. Nothing matters to them. Only the hallucinations are real. All they want is to get the next hit."

I was squirming now. "Don't you ever feel that if you could delve deep into your mind, you could discover—you could understand things about yourself that you've always wondered about? You could access everything that had been buried away, including long-lost memories, insights, and connections that had been just out of reach. Haven't you had a great idea in your dream, only to lose it when you wake up? There's so much locked away in our heads. If we could get to it, think how happy everyone could be."

"Frankly, I'm happy enough as it is," he said with a grimace. "I couldn't stand being any happier." He pointed. "We're here."

It turned out I was worried unnecessarily that we wouldn't have anything to talk about.

We followed a winding path across his pristine lawn to reach his cookie-cutter suburban house. Since he hadn't offered to help me with my bags, I dragged them along, one on rollers and the other over my shoulder. His house was too small to qualify as a McMansion, but it had typical features, such as an arched entryway flanked by two white-fluted columns. Inside, we stepped into a spacious hall with a high ceiling.

He led me past the kitchen, where his wife leaned over the sink, peeling vegetables. Her dark complexion contrasted with the bright orange of her dress. Why was she dressed up? Was it because of me? She looked at me, blushed, and went back to her task.

We sat in the living room, reminiscing about our school days in Chicago. When the conversation slowed down, Venu described his recent bathroom and kitchen renovations. Each time Venu's wife passed by, I looked at her and tried to catch her eye. She was slim, dark-skinned, but I couldn't get a good look at her face.

"Why did you choose to live so far from work?" I asked.

"The schools around here are better. On Highway 40, it doesn't take too long to get to work on a good day."

The atmosphere erupted when their daughter, Tara, returned from her friend's house. She moved like a whirling dervish, holding the hem of her pink dress and swinging it back and forth as she danced in circles.

"How old is your daughter?"

Venu asked, "How old are you?"

She finished one last circle before dropping to the floor.

"Don't lie down there," Venu snapped. "It's dirty."

She jumped to her feet, clapped her hands, and started laughing. Her hair swung as she ran up the stairs. Once she reached the top, she yelled, "Monster! Monster! Come chase me!" This sparked an exciting game of chase between dad and daughter.

Eventually, Venu showed me to the guest room, which was clean and thankfully odor-free. I spent the next hour unpacking a few essentials, ironing and hanging up my suit, and freshening up in the bathroom before Venu called me for dinner.

Dinner was awkward, to say the least. We sat around a kitchen table with a round glass top and a twisted, concrete base that looked like cheesy modern art. The adults ate in silence, broken only by Venu's occasional cough. The meal was simple yet tasty, consisting of basmati rice, fish, dal, and curried cauliflower. Tara sat in a toddler seat behind a plastic plate heaped with macaroni and cheese and carrot sticks. I tried to start a conversation with Venu's wife; he hadn't told me her name, and I felt too embarrassed to ask.

"The food is delicious," I said. "Thanks for hosting me. Has Venu talked about me?"

She nervously looked at her husband before replying, "He told me you had been friends once." Venu made a subtle clicking sound with his tongue.

She had a thick accent and was a transplant from India. Venu, on the other hand, was born and raised in the US. He must have gone back to India to marry her, which made me suspect an arranged marriage.

I looked at each of them. Neither was smiling, nor did either look at the other. I wished I had my ornate knife to cut through the tension in the room. I tried to talk to her again by asking about each dish. For every question, she offered only the barest details. I noticed her hands trembling as she spoke.

"Does Tara not like Indian food?" I asked.

Both parents looked at her as she scooped up the mac and cheese.

Venu answered, "She likes plain rice. We used to give her fish, but a fishbone got stuck in her throat. Ever since then, she won't eat fish."

As the evening ended, Venu showed me photos on his laptop. The angles and lighting made it obvious he was trying to be artistic. Tara appeared in many of the pictures, but none included his wife.

I wondered how abusive he had been to her to make her so timid and afraid. But after thinking it over, I realized that maybe my presence had scared her. Who knows what horror stories Venu told about his unhinged college friend? I can't imagine what she feared I might do to her and little Tara.

"I'm exhausted," I yawned deliberately. "I should go to bed to be fresh for the interview." We discussed plans for the next day, and he walked me back to my room. He was about to leave when I added, "Tara seems like a great kid. I guess you could almost say that she's your legacy."

In college, Venu and I were both best friends and rivals. He excelled at sports and with girls, while I consistently earned higher grades. I'm sure he thought I was pathetic, being thirty years old and still looking for a job. But that's the nature of academic life. I'm sure he was thinking, 'What did all those good grades get you?' He didn't understand me. He never understood me.

Venu paused at the door, hesitating. "I guess you could say Tara is my legacy." He gripped the door handle.

"And you would do anything to protect her, I'm sure," I said.

"Yeah. I would," he said, turning to face me. "What's your point?"

"You wonder why I do the work I do and why I'm so committed to this crazy path I've chosen," I said. "It's because I don't have a legacy like you. My work is my legacy, and I would do whatever it takes to protect it. I'm sure you can understand that."

"Good night, Narin." He left and slammed the door.

I spent the next hour reading my article in *Nature*. It caused quite a stir when it was published, and I was the first author. We submitted it before Cranston agreed to work with Cohen, which was fortunate. He might have taken credit for that work, too, if he could.

I was feeling tired. I changed clothes, lay on top of the bed covers, and imagined the upcoming conversation with McGregor. For weeks,

I studied McGregor's recent articles and memorized key details of his research. I thought he would be impressed by my knowledge of topics only tangentially related to DMTA. I also read about the multi-million-dollar glass and steel addition to the old brick-and-mortar School of Pharmacy.

Perhaps he would try to persuade me of the benefits of joining his department, such as modern labs and a strong record of securing NIH and foundation grants. We might talk about baseball (the Cardinals were off to a good start, from what I'd read) or football (go Rams). He would highlight all the good things about living in St. Louis: large parks, free museums, the St. Louis Zoo, and affordable housing. I already knew all that because I'd done my research. But I would go along with it.

We'd discuss salary and, most importantly, protected time. With an offer, I could finally move forward with my life. I would have the time and resources to gather more data, rewrite my grant proposal, and secure full funding. I would submit it to the National Institute on Drug Abuse, which had issued the RFA concerning Red Sky. Cranston wouldn't be able to sabotage me there.

I crawled under the covers. My thoughts turned to finding a place to stay in St. Louis and getting a car. It had been years since I'd driven, but I was confident it wouldn't be difficult.

I closed my eyes, but I couldn't fall asleep. Minutes passed. The pillow was too lumpy, the clock ticked loudly, and the blanket felt suffocating. I needed to relax. My thoughts drifted to Venu. He was a smart guy in college. Sometimes, he even scored higher on exams than I did. But he had deliberately chosen the path of mediocrity.

Maybe his job was less stressful, but he would never feel the joy of knowing his work made a big difference. Humanity was heading toward a better future. Millions of years of evolution shaped the human mind. It was time to improve it. DMTA was the key, but it needed to be controlled. The secret to controlling it was stored on a flash drive in my coat pocket.

That drive held the potential to change the world, and I would protect it with my life. With that thought in mind, the pillow softened, the clock quieted, and the blanket loosened its grip. Finally, I could rest and be ready for the crucial meeting in the morning.

I was nearly asleep when a noise startled me. It was very close, almost right next to me. It sounded like a knife slicing through flesh. I was trembling as I clung to the rocky ground and pushed myself upward. A warning: this is where my story gets a little strange.

A man with long black hair and a red tunic sat on a tree stump about twenty feet away. Using quick, forceful gestures, he was working on something.

It took me a moment to find my bearings. I stood on a high bluff overlooking a red desert, with the sun blazing directly overhead. I looked over the edge of the cliff. In the midday heat, the shifting sand resembled a pit of writhing snakes. I was captivated by this strange illusion and the graceful flow of slithering lines across the desert floor. They moved without any obvious pattern, yet I felt sure there was an underlying design I couldn't quite identify.

The man sitting on the stump stopped what he was doing and motioned for me to come over. In his hand was the knife—my knife. He seemed to be sharpening the blade to a fine edge, but I had a strong feeling that he was trying to wipe the red stain from the cutting edge.

"It has a few imperfections but is otherwise well-honed," he said. He had a deep, masculine voice. I couldn't identify the accent. It wasn't American, British, or Australian. He sounded like an aristocrat from a foreign country who had learned English as a sign of his worldly sophistication. I imagined that he spoke a dozen other languages as well. The voice sounded familiar, yet I couldn't place it.

He continued, "The knife you hold is a fixed blade with two sharp edges, each as fine as a razor. Its size and shape suit you well. Keep it close at hand." He handed me the knife.

A strong wind blew sand and dirt into my face. I used my hand holding the knife to shield my eyes.

"Have you ever used a knife as a weapon?" the man asked.

I shook my head.

"Now is the time to learn." He smiled. "I will only be satisfied when you can lodge that knife deep into my chest." He stood up, and I took a step back. He was tall, lean, and muscular. He looked to me like an anime version of a wandering samurai. To reinforce this impression, a sword was sheathed on his left side. A large hemp sack at his feet seemed to bulge with other battle gear. Around his neck, he wore a golden chain with an amulet that sparkled in the desert sun.

"Before we do that," I stammered, "perhaps we should introduce ourselves. Hi, I'm Narin Roy." I would have extended my right hand, but it was holding the knife.

The man chuckled. "There will be time for that later. First, let us review the basics. Place your feet shoulder-width apart and hold the knife at the ready."

I planted my feet firmly on the ground, but I wasn't sure what to do with my hands. "Like this," he guided me. "A solid forward grip with the tip facing up and outward. Never let the tip of the blade point at your torso. Always protect the most vulnerable areas—your face, neck, and torso. Make your body small. Tuck your head and relax your shoulders. Never fully extend your arm, or you risk losing your balance and control. He guided me into the correct stance.

He stepped back and examined my stance. "Remember this posture," he said. "This could save your life."

I nodded, still thinking about the swirling snakes on the desert floor.

He continued, "A knife will help you fend off an attack. It's mainly a defensive tool, but, when necessary, it can deliver a lethal strike. The secret is patience, balance, and accuracy."

"Patience, balance, and accuracy," I replied.

"Keep your eyes on your opponent. Never look away."

I nodded.

"The knife is a weapon that is only effective when wielded by a strong and agile body. Train your arms and legs until you can quickly leap, lunge, and strike."

"I will," I said cheerfully. "I promise."

Could I really train myself to become a killer? The idea excited me. I thought about all the people from my past I'd like to stab if given the chance.

"Good." He took a deep breath. "Do you have any questions?"

"Were you the one who spoke to me the other night?"

He nodded.

"Who are you?" I asked.

He looked up at the clear sky and waved his arm toward the endless dunes. "I am the prince of this desert." He took a deep breath and continued. "Once, this barren wilderness was a great kingdom, a hub of knowledge and learning, fertile beyond measure, with armies ready to die for my father, the king. But our family failed to fulfill its righteous duty and confront the evil monster hidden in the darkness. Even worse, we embraced and fostered the bloodlust of this monster when we thought it would help us destroy our enemies' armies. As a result, the monster grew over generations. Until one day—" He paused as if for an eternity. "One day long ago, when I was a child, I unwittingly opened the path for this evil creature to enter our kingdom. It killed without mercy. No man or animal was left alive." He looked over the edge of the cliff at the endless dunes. "Now ogres and banshees run wild through the ruins of what was once my beloved kingdom."

"That's all very interesting." I couldn't help but sound skeptical. "But what exactly does that have to do with me?"

He stepped forward, and his cool shadow fell over me. "Great evil threatens you. Long ago, your ancestor was my beloved and fearless teacher. I have pledged to help his descendants. Danger is near, and you stand at the crossroads of victory and defeat. I am here to prepare you for the mortal battle ahead."

"Mortal battle?" I was confused. "Are you from the Middle Ages, or maybe Middle-earth?"

He didn't respond.

"What danger?" My heart pounded. "Who's after me?" I thought about the flash drive. Did someone know what I knew? Did Cohen know?

"Threats approach from all directions," he said solemnly. "Guard against one, and another will overtake you. You must be prepared for anything and everything."

He began to walk away.

I called out to him, "Wait! I have more questions!" I had so many questions.

He looked at me with a stern expression and said, "This is not the time. Practice what I have shown you, and I will return to check on your progress."

Six

At this point, you might understandably ask, "What the hell was that all about?" I asked myself the same thing. I thought it was just a crazy dream, but I didn't have time to dwell on it. That morning, I had an interview that would determine my future.

I quickly got dressed and double-checked that my suit still looked good. We drove east into the city, shrouded in thick fog. Venu mentioned that this kind of fog was very unusual. He dropped me off near the School of Pharmacy thirty minutes before my appointment. The fog gave the campus a surreal, dreamlike quality. I thought to myself that, ironically, the dreamland of the Desert Prince appeared bright, crisp, and tangible, while here in the reality of St. Louis, the world seemed hazy and unreal.

I checked the directions on my phone. Not only was the fog obscuring my view, but there was construction everywhere. Sidewalks were closed, and workers in orange hard hats operated cranes and jackhammers, tearing up the concrete. It was a struggle to reach my destination. I had to ask for directions twice to get past all the barriers. The construction noise and the confusing path left me feeling unnerved and disoriented. A terrible migraine, one of my bad ones, was brewing deep inside my head. And, of course, I'd forgotten to bring my migraine rescue medication.

I entered the School of Pharmacy and greeted the receptionist with a smile. She gave me directions and pointed the way to the elevator. The place smelled fresh, and the noise from outside couldn't penetrate the stone building. I crossed a third-floor bridge and entered the new glass and steel addition. It was sparkling. Cohen's lab had been in the basement at the end of a dingy hall. I'd spent four miserable years there, working tirelessly. I realized how much I wanted to be in this place, with its floor-to-ceiling windows and crisp, whitewashed walls. I'd never wanted anything so badly.

I arrived at the administrative office; the brewing migraine had eased up, and things were looking better. I was still ten minutes early, so I looked at the names on the departmental roster, topped by McGregor. To me, the roster looked like the rungs of a ladder. I was ready to start climbing that ladder and work my way to the top.

There weren't any mirrors nearby, and I didn't see a bathroom, so I checked my reflection in the window to see if I still looked okay. I was straightening a few unruly strands of hair when Dr. McGregor walked past. I recognized him from his online picture, although his hair, eyebrows, and mustache were grayer than in the photo on their website. He had prominent creases across his forehead and wore square, thick-rimmed glasses perched on his pointed nose. He was lanky, moved quickly, and leaned forward as he walked. He entered the departmental office. I paused, pretending to study the posters on the bulletin board, and practiced my smile. Finally, he came out, sorting a stack of mail. I stepped in front of him. "Hello, Dr. McGregor. I'm Narin Roy."

He paused and examined me.

"I'm here for the interview," I said.

"Oh," he blurted out and continued sorting his mail. "Come with me."

"I see that Dr. Hussain is here." I walked quickly to match his long stride. "We worked together in Chicago. He's a great guy."

McGregor didn't respond.

We entered his lab, and I followed him into his office, a room filled with books, journals, and stacks of paper that seemed timeless. An old photo of smiling toddlers hung on the wall. McGregor sat at his desk across from me, took out an intimidating-looking letter opener, and opened envelopes. After an awkward silence, he set the letter opener down on the desk and looked at me.

"Narin, I reviewed your CV," he said, scowling. He picked up the CV as if weighing it in his hand and set it aside. "You've been working with Jerry for the past few years."

"Dr. Cohen says, 'Hello.'" I nodded several times. "He speaks very highly of you. That's why I applied for this position. He thought I would be a great fit for your department."

He folded his hands and sighed. "We receive many applications when an Assistant Professor position becomes available. The first thing I do when reviewing an application is cross out any fluff articles in low-impact journals. In your case, that leaves you with seven articles: the *Nature* article, which was highly cited, and the others, which frankly did not add much to our understanding of DMTA. We have seven publications and about twenty abstracts. And," he said pointedly, "you have no funded grants." McGregor gave me a long stare. "What should I make of all that? What would you make of that if you were in my position?"

I struggled to find something to say. I considered all the lines I had practiced, but none seemed appropriate here. "I don't think my CV fully captures the breadth of my work over the past four years."

"Well, it should," McGregor shot back. "That's what a CV is for."

I wanted to tell him everything about how Cohen and Cranston cheated me, but I doubted this approach would get me anywhere.

He placed his open palm on my CV and said, "I can only judge by what is in front of me." He started tapping on the document. Each tap echoed in my head. It was almost like he was summoning a migraine deep inside my brain.

"This document indicates that Jerry hasn't pushed you hard enough." McGregor's mouth twisted into a grim expression.

"Dr. Cohen has been very kind," I replied lamely. What the hell was I supposed to say to that?

"Understand this, Narin," he said. "I'm not saying that you're to blame. The job of a scientific mentor is to push you to the next level. Cohen has not done his job. It's a shame because you're an intelligent young man."

"I haven't been as successful as I had hoped," I mumbled. "There were some issues." My voice trailed off. I was sweating uncontrollably. I rubbed the palms on my pants. *Well, this suit will need to be dry-cleaned again.*

"There are always issues, Narin. When doing science, issues will always arise."

We sat silently for a moment.

"My R01 was nearly funded," I blurted out. "It received a good score. Cohen said that in other years, it would have easily been funded."

"Let me be completely honest with you," he said. "Without current funding, you can't secure a position like this."

"I believe that with protected time—" My voice trailed off. I knew I had lost. The game was over. But I needed something more. I needed a way forward. I wasn't ready for a knife in the chest yet. "So, what you're saying is that—I'm not ready for the position you have available?" I wet my lips and fought the urge to flee, to jump up and bolt out the door. I glanced at the door, my body and mind pushing me to end this.

"I have several applicants more qualified than you," he said. "If something in your CV suggested you could succeed in this position, I might consider taking a chance. But frankly, I don't see it. It's unfair to you, the other applicants, and the institution to offer you a position where you're unlikely to succeed. I'm sure you understand." He leaned back in his chair and looked up at the ceiling. I had a brief vision of myself thrusting that letter opener under his Adam's apple. The Desert Prince emphasized the importance of maintaining eye contact with your enemy. He looked at me again with pity in his eyes. "It's unfortunate you traveled here for nothing. If I weren't so busy, I would have told you not to waste your time."

I was still struggling to find something to say. "I appreciate your honesty," I said. "I do. And I'll admit I haven't been as successful as I wanted. But what should I do now, if you don't mind my asking for your advice? I mean, really, what am I supposed to do?" My voice broke, and I felt utterly humiliated. Cohen would hear all about this. And when Cohen found out, all my former lab mates would know. They would see that I'd utterly failed, and even worse, I'd broken down like a pathetic child.

McGregor stroked his mustache. "If I were in your position, I'd do another postdoc. Spend a few years with someone who will challenge you to work hard." His face brightened. "I know that Cranston at NYU has taken a very innovative approach. I spoke to him at the last Neuroscience Meeting in DC, and he asked me for suggestions for promising postdocs. You can learn a lot from him. If you'd like, I can give him a call."

"Thank you for your time," I whispered. I stood up and walked out the door without saying another word.

Finding Hussain's lab wasn't hard. An international team of graduate students worked at sterile stations. *This was what two funded R01 grants get you.*

I asked, "Is Dr. Hussain here?"

Two looked up, and one responded, "He is out of the country."

I examined the medieval mortars and pestles at the School of Pharmacy entrance and called Venu.

"Do you remember Hussain?" I was revved up and talking a mile a minute. "He's the guy who loved cooking and always made those stupid puns. You remember him?"

"Yeah, I remember. What about him?"

"He's here now. He has his own lab. I'll be hanging out with him this afternoon. You don't need to pick me up. I'll have dinner with him tonight. We have a lot to talk about."

"How did the interview go?"

"Great," I said as cheerfully as I could. "It couldn't have gone any better."

I thanked the receptionist and left.

I needed to clear my head, so I followed a group dressed in hospital scrubs as they headed north on Euclid Avenue. The fog had lifted, replaced by a muggy haze that made everything look dirty and blurry around the edges.

I crossed Forest Park Parkway into the Central West End neighborhood. For those unfamiliar, it's a trendy part of town on the western edge of St. Louis City, with restaurants, art galleries, and coffee shops mixed among apartment buildings and historic mansions. Smartly dressed businesspeople entered upscale restaurants, while sweaty but stylish exercise enthusiasts exited yoga studios. As I passed, a disheveled man with a dirty beard, knitted cap, and ragged trousers eyed me closely. I wanted to tell him, 'Hey buddy, save me a spot. I'll be joining you soon.'

I had been walking for about ten minutes when I came across a bookstore called Left Bank Books. I went inside. I've always enjoyed bookstores, and I particularly liked this one because it had the charm of a local shop while also offering an extensive selection of books. As I mentioned earlier, I dislike movies because they're just fictional stories. However, for some reason, novels feel different. It's hard to explain why, but I can become completely absorbed in a good book. After an hour of browsing, I left with *The Insulted and the Injured* by Dostoevsky.

I entered a nearby café, ordered breakfast, and spent the morning reading and sipping tea. The atmosphere was warm and inviting, with original artwork on the walls and live plants in the windows. I noticed a bulletin board announcement and made a call.

"I'm interested in the room you have for rent. Is it still available?" Although I had no idea what the future might hold, I knew I wouldn't be returning to Chicago anytime soon.

Seven

I walked down a tree-lined street toward Ms. Flowers' house. She was the one advertising the room for rent. My GPS predicted a seventeen-minute walk, but after a quick detour around some construction, I arrived in nineteen minutes. The neighborhood looked nice, although it was always hard to tell when you're in unfamiliar territory.

Her house was made of red brick, with peeling white trim and a faded red door. The concrete steps leading to the door were cracked and crumbling in places. Vines grew on one side of the house. Since there was no doorbell, I knocked as hard as I could on the heavy iron knocker. The sound shook the door and seemed to echo through the neighborhood. I waited a full minute before seeing movement through the frosted glass. Finally, the door opened slightly.

"Did you call about the room?" a voice asked from inside. The voice sounded old but was surprisingly strong.

"Yes, my name is Narin Roy. We spoke on the phone," I explained. The door closed again for a few seconds. I swallowed nervously, waiting until I heard the rattle of a chain, and then the door creaked open. Standing before me was Ms. Flowers. She was a slender woman, but I wouldn't call her frail. She had a fragile elegance. Her shoulders were draped with a gold-embroidered shawl that looked old but well-loved. 'Vintage' was the word that came to mind. Her silver hair was styled in a bun. She was definitely old, but I am a poor judge of age. Once again, 'vintage' comes to mind.

She led me into the living room and sat me down by the fireplace. Candles burned on the mantelpiece. She sat across from me on the other side of the coffee table. The room was filled with a variety of ceramic figurines, mostly of children, horses, dogs, and cats. In an alcove above the door was an arrangement of dried flowers. The candles were vanilla-scented and extremely strong. I realized my head might explode if I stayed in this room too long. I thought about asking Ms. Flowers to blow them out, but I didn't want to get into a lengthy explanation. I just wanted to finish the transaction and leave.

She probably had the same thought. She looked at me suspiciously when she opened the door. Maybe the brown man in her living room was part of some terrorist group. If you trust the news reports, you have to be cautious of outsiders. Insurgents could be planning harm to America, even in this peaceful neighborhood. Or maybe she just thought I looked shady. It helped my credibility that I was still wearing my interview suit.

She told me about the rental three blocks north of her house. Although the house was no longer well-maintained, she said I would find it comfortable enough.

"I don't fool with contracts and leases," she said. "You pay for one month at a time at the start of each month."

"That's perfect," I said.

"Here are the keys." She handed me the keys on a metal ring. "The larger one opens the front door. The other is a key to the room right at the top of the stairs. That would be your room. I can't walk you to the house because of my arthritis, but it won't be any trouble for you at all." She looked me over from head to toe. "Take a look at it. If it fits your needs, I'll expect payment on the day you move in."

"Does anyone else live there?" I asked.

Her face brightened. "There's a man who rents the room downstairs. His name is Sanquel. He's an excellent cook and always brings me treats. He's also kind enough to take care of the yard. I think you'll find him a good housemate." With that, she led me back to the entrance. "By the way, there are to be no pets," she added as I headed out the door.

Once outside, I took several deep breaths, trying to clear the vanilla scent from my nasal passages. I entered the address into my phone and followed the directions. As I got closer, I looked for the house, but all the addresses were hidden behind winding vines, overgrown branches, and ragged bushes. My phone finally pinged: "You have arrived."

It was a dilapidated house that blended in with the other old homes on that block. Thorny shrubs guarded the windows. A large, twisted gumball tree (I think they're called sweetgum trees) towered over the house, its broken branches hanging precariously overhead. The front yard sloped steeply, so I had to climb many stairs, which left me winded. I was completely out of shape—so much for preparing for mortal battle.

The front door swung open into a living room furnished with two green velvet sofas and a water-stained coffee table. A well-worn oval rug lay on the brown shag carpet. A small TV sat on a rickety stand. At the far end of the room was what I believed to be my future housemate's bedroom. I wanted to peek inside, but the door was closed.

I examined the kitchen. The linoleum bubbled and lifted in several spots, and an old computer sat on a small dining table surrounded by three mismatched chairs. I looked inside the fridge, which was surprisingly well stocked. I remembered that Ms. Flowers had mentioned the other tenant was a good cook.

I walked back through the living room and then headed upstairs. There was a small bathroom with a shower/tub combo and a shower curtain, which was oddly decorated with a stylized crouching tiger with razor-sharp fangs. Everything looked worn but clean and smelled fine. I unlocked the door to what would be my room.

Sunlight filtered through yellowed blinds, illuminating dust motes drifting through the air. The smell in the room was stale but not unpleasant. The four bare walls were painted a sickly green. A dusty ceiling fan hung overhead. A thick layer of dust covered everything—dresser, bed, and floors. I wasn't impressed with what I'd seen so far. I preferred everything to be clean and tidy. Bringing this house up to my standards would require a lot of work. I turned on the light.

That's when I noticed the sign, or at least that's how I saw it. There was something strange on the wooden floor in the middle of the room. Even though the dust was otherwise untouched, a shape had been drawn into it. It looked like two arches crossed at an angle. The shape seemed familiar, but it took me a moment to recognize what it was. It was the symbol worn by the Desert Prince on his amulet. There was no doubt about it. I believed this was a sign from him, telling me I was on the right path. Of course, part of me thought this was all nonsense. Still, it made me decide what to do. I called Ms. Flowers and told her I'd bring her a check the next day.

So, I had a place to stay. I still needed a car and a job. I decided not to dwell on it too much. If I had the support of a supernatural, otherworldly prince, everything would work out fine. A gentle breeze had blown away the last of the murky vapor, and it turned out to be a nice day. It was time to explore St. Louis. I would walk to the famous Gateway Arch and see the Mississippi River.

I locked the rental and kept walking east. I looked up some facts on my phone about the Central West End, where T.S. Eliot (before deciding to become British), Tennessee Williams, and Joseph Pulitzer had lived. It was a notable list of figures. The GPS said it was a 2-hour walk to the Arch. It didn't specify if I'd pass through unsafe neighborhoods, but I didn't care. 'Caution to the wind,' I told myself.

I headed down Lindell Blvd and just crossed Grand when I realized how hungry I was. About ten minutes later, I found a place called The Fountains on Locust. It was more of a dessert shop, but it also served real food. I took a seat at the counter and ordered chili with cornbread on the side. I finished it off with a giant double fudge sundae. The food was fantastic.

While I ate, I listened to two gruff old men grumbling about "all the damn construction downtown" and how it slows traffic, as well as how hipsters now move back into the city and turn old, rundown buildings into luxury lofts, which drives up everyone's rent. They complained about the junkie Red Skyers wandering around streets, urinating in the elevator shafts, sleeping on the sidewalks, and that the city "doesn't do nothing about it."

With my stomach full, I was feeling good, but it wasn't long before my mood darkened. McGregor probably called Cohen right after our meeting. Cohen would have already told Jason Chen. Jason, who had moved to San Francisco, always kept track of all the juicy lab gossip. By then, even the old couple from Russia, who worked scrubbing test tubes in Cohen's lab (former renowned scientists in the USSR), were lamenting my bad luck.

The hatred and anger that had always lingered nearby resurfaced once again. My mind flooded with toxic thoughts. *Cohen is a coward.* His problem was that he constantly sought validation for his ideas, which is why he turned to Cranston for support. He feared the scrutiny of our results without Cranston on board. Cranston then took the opportunity to claim all the credit for my discovery. Cohen became the third author, and I ended up with nothing!

I became so agitated that I got lost. I stood in an empty lot filled with old tires, beer bottles, and other discarded items. Nearby, there was an overpass. I needed to figure out where I was. At first, I thought I was alone, but then I noticed the others. There were three of them, and they were so still that they were easy to overlook. Two were women, and one was a man. They were just a few yards away. The women were squatting on their heels, tilting their heads toward the sky. Their eyes were wide open, as if they were witnessing something wondrous or terrifying beyond words. The man was looking directly at me. I froze.

I started to step back, initially slowly, then more quickly. I remembered the Desert Prince telling me to keep eye contact with my attacker. There was no aggression, but I wasn't taking any chances. So much for throwing 'caution to the wind.' I scolded myself for my current predicament. I only wished I had my knife. I wondered if this was the mortal battle the Desert Prince had warned me about. Then, I fell.

As I stepped back, I tripped over an old hubcap partly buried in the rubble and fell hard onto my back. A jolt of pain shot through my left palm. Looking down, I saw it was covered in blood; I had cut it on a broken beer bottle. The bleeding was heavy, so I tried to stop it with my suit jacket. Then everything went dark.

When I woke up, I was lying under the overpass. I could hear traffic passing above me. I looked at my hand; it was wrapped in clean white gauze. It throbbed, but there was no blood. What had happened? Who had cleaned and bandaged my hand? Then I saw him. As still as a statue, the man I'd seen earlier was just a few feet away, staring intently at me. I was terrified. I tried to speak, but the words got caught in my throat. He broke the silence.

"You fainted," he explained, "after you cut your hand."

He appeared young, with a few days' worth of stubble on his chin. I guessed he was in his early twenties. Lean and gaunt, he had long, messy hair. He wore a T-shirt that said, 'I got crabs at Louie's Crab Shack!' Somehow, that made him seem less intimidating.

I found my voice. "Uh, thanks," I muttered. "I owe you one." I looked at my bandaged hand. "So, where did you get the first aid supplies?" I asked, gesturing at the deserted surroundings.

He replied, "We need to be prepared for everything."

"We? Everything?" I struggled to find the right words.

"We Red Skyers. And everything means everything." He spoke to me as if I were a child.

"Tell me more about Red Sky," I asked. Although I understood the science behind Red Sky's effects on the brain, I hadn't spoken to a user until then.

"What is there to tell? It is everything," he explained. "Until it is not. When that happens — when the congruence begins to unravel — anything can happen. That's why I have bandages. I also have sutures, tourniquets, compresses, skin staples, paracord, surgical scissors, and hemostats. Oh, and duct tape. That can be surprisingly handy. One must stay vigilant.

The mortal battle comes when you least expect it. But you will learn that soon enough." With that, he walked away.

"Wait!" I called out to him. "Tell me more! I don't understand." I tried to sit up, but I felt woozy and collapsed again. I lay flat on my back for what felt like forever. When I finally managed to stand, I was alone.

I walked away feeling weak-kneed. My hands trembled, and I nearly dropped my phone as I struggled to find myself on the map. I headed to the nearest intersection to get my bearings. I stood at Locust and Broadway, just minutes from the Arch. A wave of relief and happiness swept over me when I saw the stunning metal structure gleaming in the late afternoon sun. I crossed the green lawn, passing families and vendors. Walking under the Arch felt transformative, as if I was entering a new realm. Then the euphoria disappeared, and I scolded myself for falling into magical thinking again. To stay sane, I had to focus on the real world. I couldn't get lost in fanciful thoughts of mortal battles and the universe's harmony. I needed a job.

I checked the time; it was 4:33 pm, and I still had some time to kill. I sat on the concrete stairs, pulled out the book I'd bought, and examined the cover. A long, uneven tear, held together by yellowing tape, ran down the front of the red dust jacket. Dostoevsky's name was at the top in bold white letters. Below that was the title, *The Insulted and Injured*, printed in an off-white color against a sea of red. I opened the book and resumed reading from where I'd left off.

By evening, Prince Valkovsky's scheming and cruelty started to weigh heavily on me. I went down to the river and watched the dark water and the white eddies along the shoreline. This was the mighty Mississippi River. The sunset turned the Arch into a brilliant amber, reminding me of the glowing edge of my knife.

A cool breeze started to blow, and a few Red Skyers wandered past with vacant stares and zombie-like shuffles. As the sun set, I crossed under the shimmering Arch again and called a taxi to take me to Venu's house.

Eight

Venu greeted me at the door. I had prepared a package of lies to handle the flood of questions about my interview. Venu seemed uninterested, even after I mentioned that the interview had gone better than expected. I told him I had found a place to stay and would move in the next day. That brought a smile to his face. He noticed my bandaged hand but didn't ask any questions when I explained that I had tripped on uneven pavement and cut myself.

We ate egg salad sandwiches, loaded with mayo, at the kitchen table.

I asked, "How was your day?"

"I had a shift working at the pharmacy in the ER today," Venu said. "I saw a Red Skyer who had tried to kill herself. Her mom said she was a valedictorian in high school three years ago, but she started using Red Sky in college to cope with stress. When she was brought in, both her wrists were covered with blood-soaked bandages." His gaze fell on my bandaged hand. "She was so pale. She looked like a ghost. They had to pump her full of antipsychotics to keep her calm enough for the blood transfusion."

"Did she say what happened?" I felt sick.

"She didn't say much of anything at first," he replied between bites. "But the doctors kept pestering her until she said what they wanted to hear. They wanted her to describe how everything felt unreal. They aimed to fit her into that neat little diagnosis called 'DMTA-induced psychosis,' what everyone else calls 'the switch.' Eventually, she mumbled something about maybe everything feeling unreal. That was enough to get her admitted with more complexity, which allows them to bill more for her hospital stay."

As I pictured the pale young girl with blood-stained bandages wrapped around her wrists, I tried to eat but couldn't. I realized I was trembling. I muttered something about needing to pack and went back to my room. Once there, I collapsed onto the bed and reflected on my day.

Within twenty-four hours, I lost all hope of pursuing a career in academic pharmacology. My years of hard work and sacrifice seemed wasted. My dreams of making the world a happier, smarter, and more creative place appeared to vanish into thin air. Then I had that encounter with the Red Skyers. What was that about? I remembered meeting the Desert Prince in my dreams the night before and being told I must prepare for a mortal battle. At least I had found a place to live, even though it was a dump. At that moment, I desperately needed a hot, steaming shower.

In the bathroom, I removed the gauze bandage from my hand. I wanted to clean and carefully examine the wound. I wasn't sure if a homeless druggie had provided proper medical care. But my doubts were unfounded. The wound was deep and required several stitches. The sutures were neatly tied, and a thin layer of antibiotic ointment had even been applied.

After my shower, I brushed my teeth and changed into my sleeping clothes. I placed my poor, abused suit in a bag to send to the cleaners. Before going to bed, there was a call I couldn't put off any longer.

A child answered, "Uh, hello?"

"May I speak with your father?" I asked. I heard him yell for his father.

"This is Jason Chen."

"Your kid sounds all grown-up," I said.

"Narin, is that you? Oh, my God. How are you? How did the interview go?"

I didn't buy it. He knew everything but acted like he didn't know, so I went along with it. "McGregor wouldn't commit," I replied.

"Bummer. I was afraid of that."

"But I called to talk to you about something else, something important," I said.

"Sure. Anything. What's up?" I could hear the hesitancy in his voice.

"We've been working with DMTA for a long time—longer than nearly anyone else. Have you noticed any health effects?"

"What kind of health effects?" Jason asked.

"Hallucinations, weird dreams, or anything like that?"

"Listen, Narin. Sometimes I catch something out of the corner of my eye, and I look, but there's nothing there, and I think, 'Wow, I must be hallucinating.' Somehow, my DMTA receptors got triggered, and now my brain's all messed up. But then I remember that any potential DMTA exposure I might have had in the lab is tiny compared to what a Red Skyer takes with just one hit. There's no way it could have any effect."

"But maybe there's a cumulative effect," I pressed on. "I've had some bizarre experiences lately. Just yesterday, I had this intensely vivid dream. It felt like it was really happening. I'm getting spooked."

"Weird things happen to all of us," he replied. "You're just noticing it because—because things suck for you right now. Listen to me, Narin. Cohen screwed you. And those bastards at the NIH screwed you also. You should've been funded. Everyone knows that. Cohen knows that. I know that. But you can't let this little setback get you down. You're going to succeed. Do you understand me?"

When I didn't reply, he kept going. "You've worked too damn hard to give up now. You've got to fight for what's rightfully yours. You need to reapply and get funded. Listen, I was talking to one of my colleagues here. He told me that people who have a near miss with the NIH tend to do better over the next ten years, publishing more high-impact articles than those who barely make the cutoff. That's because this kind of setback can ignite a fire under your ass and make you stronger. That's what's going to happen to you."

"You're probably right," I replied. "Also, do you happen to know if anything is available in your lab or any other labs at UCSF?" I'm sure he saw that coming.

"They just announced a hiring freeze," he replied. "It's a bad time."

"What about a postdoc position? I'm desperate. I'll even wash test tubes if that's all you have. I need a job right now."

"Here's what I'll do: I'll look into it and get back to you. I promise."

I hung up. What a lucky bastard. He's got it made. I've got nothing.

I rummaged through my bag and pulled the knife out of the wooden box. It was shining as if it had just come from the forge. Using my thumbnail, I scraped off the grime from the space between the wooden handle and the blade. It was time to practice my stance.

I placed my feet shoulder-width apart and closed my eyes. I tried to remember what I needed to do. I adjusted my stance until everything was aligned and my muscles relaxed. I stood still, taking slow, deep breaths. Then something unexpected happened: I saw a murder.

I'm talking about the murder my grandfather told me about—the one supposedly committed with the very knife I was holding. Before me stood a young woman in a lavender sari. Her heart-shaped, youthful face was framed by dark eyes wide with fear. She cried out as an older man, probably her father, wielded a knife at a trembling, pale, scholarly-looking man who knelt on the floor before him. The cowering man couldn't have been more than twenty years old.

As I watched, the father, wearing a morbid grin, closely scrutinized the face of the terrorized youngster. Then, with calm deliberation and a groan of pleasure, he slit the man's throat. The woman screamed and ran toward her lover, now collapsed on the floor, blood pooling beneath him. As she caressed his bleeding body, her father approached her from behind and, with a single stroke, sliced her throat so savagely that her head was nearly severed from her body. She fell heavily onto her young lover, joining him in death.

You might think I would be horrified by this terrifying vision, but all I felt at that moment was exhilaration. I had life and death in my hand. Why do people chase safety, stability, prestige, and power when all of these can be ripped apart in a single moment? Happiness is fragile; it can easily be shattered.

Then the scene shifted. My grandmother and I played Snakes and Ladders at my parents' kitchen table while my mother worked at the counter. We listened to an old scratchy record, and the songs were in Bengali, but I couldn't understand what they meant. I asked my grandmother to translate.

"Oh, this is a beautiful devotional song," she said, nodding in appreciation. She then translated the poetic Bengali into a more conversational form, which I understood. "Oh, Mother, how I long to be with you. A moment without you is like an eternity of suffering. Please take me now so I can experience your magnificent bliss. This joyless life has nothing to offer me."

"What is the hurry?" Mom asked. "Life is short enough as it is." Her voice quivered as she scrubbed the counter.

"What do you mean, Mina?" Thakur-ma asked.

"Why should you be in such a hurry to die? We should appreciate the life we have." I could tell from Mom's tone and body language that she was pissed.

"When your heart is with God, you lose all desire for worldly things," Thakur-ma said, rolling the dice. "Your mother, the good woman she was, understood what I meant." She moved her piece. "When her body wore out, she chose to be with God rather than cling to life."

"Do not speak of my mother," Mom yelled. Her face was flushed, and the veins on her forehead bulged. She dropped her dishrag and left it on the floor.

"Mina, I only have kind words to say about your mother," my grandmother said. "It is no secret that when she was in severe pain, the cancer doctor gave her medication to relieve it. She took the pills, one after another, until the pain was gone. There's nothing wrong with that. Her heart was with God. There was nothing left for her in this world."

My mother jabbed a finger at her. "What about Narin and me? Are we nothing?" The veins on her forehead twisted like snakes.

"You weren't there," my grandmother said, shaking her head.

I rolled the dice.

Mom stormed out of the room.

"How unfortunate for you, Narin. You have fallen into the mouth of a snake." Thakur-ma smiled wryly. "Down you go."

I had tightened my grip on the knife, or perhaps I should say, the knife had tightened its grip on me. My knuckles turned white, and my heart pounded in my chest.

The phone rang, interrupting the moment. I was snapped back to reality.

Two rings, three rings, four rings.

I placed the knife on the bedside table, unsure if my fingers would obey my command to let go. The knife hit with a thud.

It was my sister.

"I didn't get the job," I said.

"When are you coming home?" she asked.

"I'm not in a rush. I can't go back to my old job." I sighed. "I'll call that guy from Harvester tomorrow to see what's available. Right now, I have no idea what I'll do. I found a room I can rent for a month. I'll stay here and see what happens."

"How's your mood holding up?"

She hadn't cared before, but now she cared.

"I'm so used to being buried in my work. All this freedom gives me too much time to think. I keep replaying recent events and wondering what I could've done differently so I wouldn't be in this humiliating position now."

"Why do you want to do research anyway?" she asked. "Maybe you should just get a regular job."

I didn't like how that question was asked.

"Because I think it's important," I replied sharply, twirling the knife on the bedside table. "I believe Cohen and I were onto something. All I need is the time and resources to explore the ideas I proposed in my R01. Whoever unlocks the secrets of DMTA will undoubtedly win a Nobel Prize. There's no question about that."

She chuckled. "You have really high expectations."

"Shouldn't I?" I was getting pissed off now. She knew how to push my buttons as well as I knew how to push hers.

"You should," she said nonchalantly, "but things might turn out differently than you expect. You're lucky to have some viable options. Most people aren't so fortunate."

"You may not understand my situation," I yelled. "I don't have any viable options."

"Okay, Narin." Deepa yawned. "You do whatever you think will make you happy."

"I know the people who got funded. I read their grants." The knife fell to the ground, and I picked it up. "Their applications sucked compared to mine. I just got cheated. That's the bottom line, end of story. One of the reviewers was a complete idiot. He had no idea what we were doing." I wasn't sure why I was yelling at my sister, but I needed to vent to someone. "The person in charge of reviewing the grant, Dr. Cranston, had reason to make sure I didn't get funded. The entire system was rigged to prevent me from succeeding."

Deepa paused for a moment, then said, "You're yelling at the wrong person." She yawned once more. "Good night." She hung up.

I held the knife between two fingers. *That's one hell of a knife.* But what was it trying to tell me? The things it showed me weren't real. At least, they weren't real memories of mine. The first vision was of the murder the knife committed. The second was a conversation between my grandmother and my mom that never actually happened, at least not in that way. By the time Thakur-ma moved to the States, she had advanced dementia and couldn't hold a coherent conversation. Also, I had only played Snakes and Ladders with her as a child visiting India, not as an adult in my parents' house. The meaning was lost on me. Maybe the knife was showing me different kinds of cruelty, both the horrific and the everyday. Who knows?

I stood up and assumed my stance. Once I was in position, my entire body relaxed. I waited for another vision, but nothing appeared. I stayed like that for a long time, expecting my muscles to rebel. But nothing happened. It seemed I could hold that posture forever. To remain alert, I imagined a snake hiding in the tall grass, watching me, ready to strike, opening and closing its powerful jaws, waiting for a moment of weakness to sink its fangs into me.

Nine

I felt more hopeful the next day. I had added another name to the list of people I wanted to leave in the dustbin of history—McGregor. Once I succeeded in creating a drug that would change the world as we know it, McGregor and all those idiots would regret what they had done to me. I stretched my arms high above my head and arched my back.

I prepared for the day and packed my modest belongings into my bag. I sifted through it to find Ian's card and dialed his number. He answered on the third ring. "Hi Ian, this is Narin Roy. We met in Chicago. I'm just calling to check in with you." I folded my arms stiffly around myself. "I wanted to learn more about what we discussed the other day—about Harvester and the opportunities that might be available to me."

"What are you doing right now?" Ian asked cheerfully. "Let's grab some drinks and talk."

"That sounds, uh, fine. Where would you like to meet?"

"Tell me where you are, and I'll come pick you up."

I wasn't sure if I was ready to talk to him, but I decided there was no better time than the present.

I hung up and went downstairs. I found Venu's wife in the kitchen. Tara sat at the kitchen table, drawing on a large sheet of black construction paper.

I thought Venu's wife would be more relaxed without Venu around. Back then, I believed she was afraid of Venu, not me. She looked angry that morning and shot a harsh glare at me when I entered the kitchen. I had no idea why she was mad at me; I hadn't done anything to upset her.

I politely asked if I could have some tea and sat next to Tara at the kitchen table.

Venu's wife (I still didn't know her name) warmed the water and poured a cup for me without looking at me or saying a word. She was so thin—unhealthily so. That morning, her shabby dress made her look like a housemaid. If she was going to ignore me, I decided to ignore her too. Instead, I watched Tara apply gold stars to the construction paper.

Soon, the silence grew uncomfortable.

"Venu must be at work," I said. "He told me that he's swamped."

She filled the sink with hot water and put in a few drops of soap.

I told Tara, "I love your drawing."

"I'm making it for you, Uncle Narin."

I've always liked how Indian kids call adults 'Uncles' and 'Aunties,' as if we're all part of one big happy family.

Venu's wife cleared the dishes from the counter and placed them in the sink.

"So, how long have you been in the United States?" I asked. I tapped my fingers on the table and took a sip of my tea.

"Five years," she answered.

"That's a long time. Do you like it here?"

She fumbled with the soapy dishes. "I do."

"Don't you miss your family?" I asked, leaning back in my chair. "It must be tough being so far from your loved ones. I'd definitely feel lonely."

She turned, gave me a cold glare, and exhaled sharply. "My family is here," she snapped.

I nodded, saying, "I guess." I held out the empty teacup, which she snatched from me and dropped into the soapy water.

"I'm almost finished, Uncle Narin." She set the last star in place. "There, it's all yours."

The golden stars outlined the shape of a giant man with arms and legs spread across the night sky. Two emerald stars represented his eyes.

"I call him 'The Man Who Lives in the Sky,'" she said, smiling as she handed me the picture.

"Thank you, Tara. It's beautiful." It's interesting how a child's instinct is always to be kind to someone, even when the parents are hateful.

I had a few extra minutes, so I put Tara's picture away, went outside, and waited for Ian. He said we were going out for drinks. I told myself: (1) I didn't want to work for Harvester, BUT (2) if I didn't get this job, I'd have no choice but to thrust my knife into my chest.

This is a good place to share what I knew about Harvester at the time. Their claim to fame was extracting plants used in traditional medicine from the Amazon rainforest and turning them into pharmaceuticals. They weren't the largest pharmaceutical company, but they had done quite well since their founding twenty years earlier.

That is, until three years ago, when lawyers representing several indigenous tribes in South America filed a lawsuit against Harvester. The lawsuit claimed that Harvester had used threats and violence (through its influence with local governments) to coerce the indigenous tribes into revealing their knowledge of medicinal plants. Shocking news surfaced about the harsh tactics Harvester had employed to obtain its blockbuster pharmaceuticals.

As the trial progressed, more reports of exploitation and brutality emerged. I didn't follow this closely, but even I heard about the negative press. If you divide the world into oppressors and the oppressed, then Harvester (at least in the court of public opinion) was clearly on the oppressor's side. Their shareholder meetings turned into public spectacle, where 'activist shareholders' waved banners and shouted at the Board of Directors, demanding justice for Indigenous Peoples.

Now, this is where the story gets interesting. Harvester pulled Savycha, a difficult-to-grow plant native to the Andes, out of the jungle in the early 1990s. It had been used in shaman rituals for centuries. With the arrival of the Spanish, it became part of certain Catholic ceremonies. In some tribes, it is still used as a panacea and love potion.

Harvester first isolated DMTA from Savycha around 1993. I remember reading about it in high school. They aimed to develop a non-opioid painkiller, but problematic side effects like hallucinations and euphoria ended the project before clinical trials. It wasn't until 1998 that it appeared as a street drug, Red Sky. No one knows precisely how DMTA went from being a trace chemical in an obscure Andean plant to the most popular recreational drug worldwide, but that path seems to go right through Harvester. Of course, they denied any responsibility for the epidemic. Given the shady connection between Harvester and Red Sky, I was shocked when Ian told me they're still interested in developing DMTA into a potential therapy.

By 2000, Red Sky was everywhere. It wasn't hard to produce, and as a pill, it was easy to transport and consume. No one saw the downside until the 'switch' was described, initially as case reports in 1998, then as a full-blown epidemic by 2001. Hundreds of otherwise ordinary, healthy people committed suicide or perpetrated murder-suicides. And as I've discussed before, these were gruesome affairs with blood and guts strewn everywhere.

In 2000, the team at UCSF, where Jason Chen now works, discovered two DMTA receptors in the brain, creatively named Type I and Type II. Dr. Eugene, who led that team, will undoubtedly win a Nobel Prize for this work. There has been strong interest in understanding the roles of these receptors and their connection to DMTA's positive and negative behavioral effects.

Cranston was the first to show, using a rat model, that stimulating the Type I receptor with DMTA causes euphoria and enhances cognition. Rats given DMTA performed better in visual-spatial tasks, such as mazes. They even learned a simple pattern code, pressing three buttons in the correct sequence to access a stash of DMTA. Rats treated with DMTA consistently preferred accessing DMTA over food or even a female rat in heat.

Our group, led by Cohen, worked with a mouse model of the Type II receptor. In one of my earliest experiments, I used a mouse strain that did not express the Type I receptor. In other words, these mice only had Type II receptors. You might wonder why we used mice instead of a species more similar to humans. When it comes to the brain, mice are remarkably similar to humans. They're inexpensive, docile, and easy to manipulate.

One day, I administered a low dose of DMTA to one of these mice. I stepped away for a few minutes, and when I came back, he had disemboweled his two littermates and gnawed off one of his hind paws. No one in the lab would get close to him. We ended up suffocating him with nitrogen gas. After conducting several more experiments, which produced similar disturbing results, we published our preliminary findings in *Nature*. It seemed that stimulation of the Type II receptor was responsible for the switch (DMTA-induced psychosis).

That's when Cohen and Cranston began their collaboration in late 2004. All my data was sent to NYU so Cranston's lab could "verify my results." Four months later, Cranston's group published the paper that established the Cranston-Cohen model as accepted dogma without even acknowledging my contributions. In short, activating the Type I receptor brings you unicorns and rainbows, while activating the Type II receptor results in disembowelment and death. After the paper was published, everyone was talking about it, and the race was on to develop a drug that could block the Type II receptor.

That's what I've spent the last 18 months working on. And guess what—I found it! I discovered an off-the-shelf drug that's been around forever and binds the Type II receptor without causing psychotic behavior. Then I had the chemistry department make a few simple tweaks, and it became an irreversible blocker of the Type II receptor. But by the time I figured all this out, Cohen was on my shit list, so I didn't tell him what I'd found. I kept the entire process secret, using one lab book for my private use and another for Cohen to see.

My R01 proposal would have been stronger if I had included this work. Honestly, I would have easily received funding. But I didn't trust Cohen, Cranston, or anyone else. I kept my secret to myself, hoping to publish it as my own work later, once I had a lab up and running. So much for that plan.

Ian pulled up in a late-model red Porsche that gleamed like a blazing fire in the morning sun. I don't know much about cars, but I could tell this one was expensive. He got out and watched as I approached. His blond hair looked even more plume-like than the first time we met. I wondered how much time and product he used to style it that way.

I felt self-conscious and wondered if my fly was unzipped or if there was something else about my appearance that was just as embarrassing. I looked like I'd rolled out of bed and thrown on my clothes from the night before, because that's exactly what I'd done. My suit still had blood stains, so I couldn't really wear it for this interview.

I hadn't shaved, but to my credit, I'd brushed my teeth. My hair isn't as curly as my sister's; it's more like a shaggy dog's. I wash it daily but rarely comb it. I shook my head, thinking this might make it look less unkempt. Fat chance.

Ian took my hand and shook it warmly. "I'm so glad we could meet. I was getting worried you wouldn't call."

I got in and sat down in the black leather passenger seat.

"Your dad seems to be doing okay," Ian said. "I always liked him. He was always so calm and wise."

The middle console lit up, and a miniature version of the car appeared on the street map. "My parents are in India," I said.

"If I'd known David Goldblum would be at the funeral, I would have brought my copy of *The Value of Nothingness* for him to sign," Ian added. You'll remember that David Goldblum was the decrepit old friend of my father who wanted me to have tea with him.

"Do you know Professor Goldblum?" I asked.

"I know him only by reputation," he replied. "I'd taken an interest in him at one point."

"Why?"

He muttered something I couldn't understand, mentioning Joyti, but his words didn't form a clear sentence. I nodded. I'd heard of Goldblum's famous book before; it was popular in the 1960s and 1970s, but I didn't know much else. I'd always assumed it was a self-help book.

"So, how did your interview go?" he asked.

"I'm considering the offer." I looked out the window. I wasn't a good liar. I started pulling at my hair to help relieve the anxiety. "I'm not sure I want to move right now. My family wants me to stay in Chicago. Besides, an academic career has its drawbacks."

"Chicago's a good town, but St. Louis isn't too bad," he said. "It's got the same problems as any medium-sized Midwestern city: annoying blowhards in the suburbs and annoying hipsters downtown."

"I went downtown yesterday," I said. "There are a lot of Red Skyers walking around."

Ian nodded. "That's a relatively recent development," he said.

"Did you grow up around here?"

"Yes, I did." Ian smiled. "Are you going to ask me the quintessential St. Louis question?"

"What's that?"

"Where did you go to high school?"

I didn't get it. "Why would anyone care about that?"

"In this town, it matters," Ian said. "You know everything you need to know about a person from the high school they attended. You know their social rank, politics, where they will end up, pretty much everything."

I looked down at my hands. It struck me as very sad that high school can determine one's destiny. *I'm terrible at this casual banter.* I tugged harder at my hair. I just wished for it to be over soon.

I asked, "What high school did you go to?"

Ian chuckled. "A small private school called John Burroughs."

I nodded, unsure of what to make of this information. "So, what does that say about you?"

He replied, "It says that I'm either smart or come from money."

"So, which one is it?"

He didn't respond, so I moved on. "How long have you been with Harvester?" If I were to join Harvester, I needed to know as much as possible.

"Four years now. And that's my father's doing," Ian said, with a twitch of his mouth. "He and Maru's father — the CEO of Harvester, Sanjay Chandra — were college buddies. My dad's on the Harvester Board of Directors."

"It helps to know the right people," I said.

"I have a question for you," Ian said. "Did you grow up in Chicago?"

"Since I was three, yeah."

How does a Chicagoan decide whether to root for the Cubs or the White Sox? Or do they root for both?

"I'm not a big baseball fan, but I grew up around Cubs fans, so I guess I'm a Cubs fan too."

Ian nodded. "That's unfortunate." He looked at me gravely. "Don't worry. Your secret's safe with me."

Every Chicagoan knows that the Cubs' last World Series win was in 1908. Since then, the Cardinals have won the title nine times. I don't understand why St. Louisans insist there's a rivalry.

We arrived at a cozy corner restaurant in a residential neighborhood. I'm trying to remember the name now, but I can't quite recall. It's strange; everything else is so clear.

I figured Ian chose this place to keep things casual. He said we were going for drinks, and maybe he thought that if he got me drunk, I would sign my life away without a second thought.

I saw Maru through the window, watching us. His oversized glasses made his face look small. Ian directed me toward where Maru sat. I gave him a friendly nod, but he scowled and looked away. The place wasn't crowded, just a few couples eating large, meat-filled sandwiches. Ian must have been a regular, judging by his chat with the owner, Hal, a big man with greasy black hair and a loud voice. Ian didn't order anything, but Hal seemed to know what he wanted. He set down a bottle of Scotch and three shot glasses.

"Only the top-shelf stuff for you and your friends," Hal said with a smile as he moved back to the kitchen.

Ian poured drinks. "You've met Maru. He's the Vice President of Drug Development at Harvester."

I nodded toward Maru. "I didn't realize you'd be here," I said in as friendly a tone as I could manage.

"I have to be here," he replied with an exaggerated sigh. "I'm here to protect the interests of Harvester." He turned his chair away from me as if our conversation held no interest for him. *What's up with this guy? Why does he dislike me so much?*

"Are you guys hungry?" Ian asked. "The buffalo wings here are incredible." Maru didn't reply, and I shook my head. I was hungry but hated eating in front of strangers, especially something as messy as buffalo wings.

"Well, let's get down to business then," Ian said. "I'll start at the beginning."

Ian waited for us to finish our drinks. If this were a job interview, it would be the strangest one I'd ever experienced. I threw back the shot and swallowed it in one gulp, just like they do in movies. My throat burned, and I struggled not to gag. I could have asked for a glass of ice water, but I was too scared. I didn't want them to think I was a lightweight. I swallowed hard and cleared my throat.

Ian continued, "As you may know, Harvester has traditionally extracted and commodified naturally occurring compounds that show potential as pharmaceuticals. The rainforests are full of untapped therapeutic potential."

I wiped my eyes. "You mean compounds that the pharmaceutical industry hasn't tapped into. I've read about the lawsuits." I wasn't sure why I brought this up. I was risking pissing off Ian. However, I thought showing them that I knew Harvester's dark side would give me leverage in the upcoming negotiations.

Ian leaned back and nodded. "As the lead attorney at Harvester, that's my department," he said dryly. "We're not afraid of some crusading lawyer, and that's not why you're here. So, one thing we pulled out of the jungle is Savycha."

"I know that whole story," I replied. *Let's get to the point.*

Ian's eyes sparkled as he said, "You also know that DMTA is a very special compound."

"Have you ever tried it?" Maru asked, cutting in. He shot me a sideways glance. Was this a trick? Maybe he wanted me to admit I was a Red Skyer so he'd have an excuse not to hire me.

I coughed again to clear my throat. "I don't use illegal drugs." That seemed like a safe response.

Maru leaned forward and whispered, "It's amazing." I wondered if he was trying to be friendly or give off the 'cool kid' vibe. Either way, it wasn't working. I always hated it when skinny, nerdy Indian kids tried to act cool; they just ended up looking like idiots.

Maru continued, "It's like getting hit by a freight train. The sky turns a deep, blood-red color."

"Sounds wonderful," I said.

"There's nothing in the world like it," Maru whispered. He took another sip and slammed his glass on the table. "But I can understand why some people are too scared to try it."

"They're scared of it because it can turn you into a bloody corpse without warning."

He shrugged and gave me a sinister smile.

I hated Maru's small, unkind face and his oversized glasses. "I'm curious, Maru. What high school did you go to?"

He looked away without saying anything. Ian fought back a laugh.

I decided not to waste my time with Maru. I told Ian, "When you told me Harvester was interested in DMTA, I was surprised. I didn't think anyone wanted to touch it because of all the negative press."

How much should I tell them? Should I reveal that I had the formula for a Type II receptor blocker on a flash drive in my right front pocket at that moment? That was my ace in the hole, but I didn't want to show all my cards.

Maru raised his hand. "Until you sign a confidentiality agreement, you don't get to know anything else."

"We're very interested in DMTA," Ian said. He refilled his shot glass and took a drink. "We've got some promising data. It's preliminary, but it shows potential. If it's a dead end, we'd rather know now and avoid wasting time and money."

I took a deep breath. "I've felt for a long time that DMTA has a lot of therapeutic potential. However, aside from the legal issues and the switch problem, it's also a naturally occurring compound. How is Harvester going to patent it?"

Ian shook his head. "Again, that's my department. And Harvester isn't just any drug company. We have a long history of working with naturally occurring compounds." He leaned forward. "Maru and I want to add you to the team working on our DMTA project. The project leader, Morey Whitely, can give you all the details about the project. In fact, he's the one who suggested we meet with you. He's been following the work of your former mentor, Dr. Cohen, and thinks you might have valuable insights we can use."

Ian's sky-blue eyes sharpened, fixing me in place. "We believe we might have a valuable use for DMTA. However, we first need more information and are prepared to invest whatever resources are necessary to obtain it. We're reaching out to you because Morey believes you can help us. And I think we can help you too. Does this sound like something that might interest you?" Ian moved closer and whispered in my ear. "Win-win, you might say." His breath on my ear sent a cold chill through me. It made my insides feel like ice. They acted less like corporate suits and more like gangsters negotiating a shady backroom deal.

I leaned back and looked at Maru, who was still staring out the window.

"I'm interested, but I need to learn more," I said.

Ian smiled. "Come by Harvester tomorrow and meet Morey." Ian set the Porsche key on the table. "Take the car. You'll need one to get around. Maru will give me a ride back to Harvester. I'll shoot you a text with the details."

What the hell? Was this Ian's car, or did it belong to Harvester? Was it a gift or a loan? Either way, I was excited to have a car. And what a car it was! It shined like a fucking ruby.

I sat inside the Porsche, inhaling the smell of leather. I wasn't sure if I liked it, but it smelled like money. I mentally reviewed a list of tasks I needed to complete that day: pick up my belongings from Venu's house, have my suit cleaned for my visit to Harvester, buy some essentials, drop off my check to Ms. Flowers, and move into my new place.

When I returned to Venu's house, his wife opened the door and gave me a slight nod. I gathered my belongings and left, thanking Venu's wife for her hospitality. I felt proud of myself for being polite despite their hostility.

Next, I visited Express Dry Cleaners, which told me I could pick up my suit after 5 pm. Cleaning off the blood stains would cost extra. I stopped by Target to pick up a few essentials. I walked out with bed sheets, a decent comforter, a hot pot for making tea, tea bags, an insulated mug, a multi-pack of gray athletic socks, a pair of black dress socks, and some navy-blue Hanes boxers. Oh, and some honey-mustard pretzel snacks and Diet Coke. I also bought a broom and dustpan combo, some disinfecting wipes, shower gel, and bandages for my injured hand. My life felt complete. I ate half the bag of pretzels on the way to Ms. Flowers' house.

I dropped off the check to Ms. Flowers that afternoon. We exchanged a few pleasantries on the porch. She sniffed when she saw the Porsche parked out front and sent me on my way with instructions to "say hello to Sanquel" and "remember, no pets." I arrived at the rental just before 5 pm and wondered if Sanquel would be around. I prepared myself for social contact, but I wasn't sure I had the energy to meet my new housemate. However, my worries were unfounded; the house was empty. I breathed a sigh of relief.

You might find it odd that I agreed to stay at Ms. Flowers' place without meeting my future housemate first. In college, I had roommates thrown into my path, and aside from Venu, it rarely worked out. But I needed a place to stay, and the place was large enough for us to maintain our distance. If it didn't work, I figured I'd move.

I spent the evening cleaning my room. I opened the windows to air out the space and wiped down all the surfaces. With a firm sweep, I cleaned the two curved marks on the floor. The Desert Prince had made his point. After some elbow grease, my room looked livable. Would Ms. Flowers let me paint the wall? The murky green was unsettling. The dry cleaners had come through for me, bloodstains and all. I hung the freshly laundered suit in my new closet. I decided to go to bed early since it had been a very long day, and I expected another doozy the next day.

Ten

Harvester was comprised of five interconnected, gray-tinted glass buildings situated on a large campus in West County. West County is an area with several cities west of St. Louis City's official limits. This region is wealthy, predominantly white, and features numerous plastic surgery clinics, LASIK eye surgery centers, and Starbucks. It's basically the suburbs. I parked at the main building and entered a spacious lobby with a black marble floor. It was eerily silent for such a large space, and the acoustics seemed designed to absorb any sound.

A pretty girl was typing at a computer behind the reception desk. Her auburn hair was pulled back into a high ponytail, and she wore a black sweater with a matching short skirt. She noticed me entering but kept typing until I reached the desk. She looked up with a smile. It was a fake smile, but she pulled it off well enough. I stood in a beam of sunlight from the overhead windows and told her I had an appointment with Ian Blair. She studied her screen for a moment and then pressed a button.

"Elevator Two will take you up to the top level," she said matter-of-factly. "Have a nice day."

I tried to see my reflection in the black marble while waiting for the elevator. The surface was shiny enough to reflect light, but I couldn't see myself. *Why am I always trying to look at my reflection?*

The elevator slowly ascended. The second floor was labeled "Gallery and Conference Rooms." Floors three through six housed all the lab spaces. I'm still not sure what the seventh and eighth floors are used for. I guess the support staff works there. The ninth floor, my destination, was the executive suite.

I stepped onto a plush crimson carpet. Opposite the elevator, a grand grandfather clock stood, its dark wood adorned with battered angel faces. Each one bore a blissful smile and eerily vacant eyes. I crept down the long hall. Ian hadn't given me any instructions besides asking for him in the lobby.

I tried to find out which office belonged to him. The doors were all stained black and had large silver knobs. None of them had names or room numbers. I heard a muffled shriek. The shrieks grew louder as the grandfather clock struck the hour. I wondered what was causing that noise. Then Ian stepped out of the room right behind me, startling me.

"Have you ever noticed how shit stains everything it touches?" He brushed his sleeve and extended his hand. That wasn't the 'hello' I'd expected. I hesitantly took his hand. Today, his handshake was neither warm nor friendly. It was very businesslike. He seemed distracted. His eyes darted around, and his brow was furrowed. His white-blond hair was still impeccable, but the top was combed down today, making it less plume-like.

Inside his office, a tall, slender bird, black from beak to talons except for its cold, yellow eyes, flapped its wings and shrieked when I entered. I believe it was a raven, but I'm not sure. It was the source of the shrieking. The bird opened and closed its long, curved beak, perhaps wondering how tasty my eyeballs would be. It didn't look friendly. The cage rattled and shook until the bird settled, watching my every move as its sharp, powerful talons clutched the wooden perch.

"It was a gift from Maru." Ian gestured toward the bird. "He likes to torture me."

With a flourish, Ian grabbed a sheaf of paper from his desk and flipped to the last page. "You need to sign this confidentiality agreement before we can get started," he said, setting the document down and pointing. "Please sign here, directly above your name," he commanded.

I flipped back and read the first paragraph on the first page. It was boilerplate, with most paragraphs starting with "The Disclosing Party shall not..." Ian lingered nearby. I didn't bother reading the rest. I'd have to sign it to find out what they wanted from me. I flipped to the end and signed the document. Ian studied my signature for a surprisingly long time and smiled. "Let's go meet Morey," he said.

He led me down the hall, speaking over his shoulder. "Morey runs the Neuroscience Division. He's in charge of the DMTA project. He'll tell you everything you need to know." We passed the grandfather clock, and I commented on how creepy the etched angel faces looked.

Ian laughed. "It belongs to the elder Chandra, Maru's father, the CEO. He brought it back from South America during one of his first trips there. He's quite fond of it. He calls it the living heartbeat of Harvester." Ian gently caressed the cheek of a cherub with a gaping mouth—an odd gesture. I realized then that he couldn't help but be creepy. Still, he could get away with it because of his good looks and charisma. I'd be arrested and locked up if I were caught fondling a grandfather clock.

We stopped, and Ian knocked on the door. It seemed odd that they didn't have nameplates on their doors. I guess if you worked here, you'd know where everyone's office was.

"He must be out," Ian said as he pushed open the door and stepped inside. I glanced back at the grandfather clock. When we passed it, I got a bad vibe. I wondered if it was haunted, but then I scolded myself for thinking such a stupid thought. I followed Ian in.

The shelves were filled with pharmacology books, many of which, judging by their worn spines, dated back to before I was born. Ian went directly to a framed photo on Morey's desk. It seemed odd that it faced outward, as if it was meant to be seen by others, not by Morey. Ian paused to look at it, then handed it to me. He peered over my shoulder. "That's Morey's daughter," Ian said. "What do you think?"

That was the first time I saw Sophie. She was standing on what looked like a ski resort slope. The picture showed two snowy peaks under a bright blue sky. In the foreground, Sophie wore a white parka and a black cap pulled down over her dark, flowing hair. Tiny brown freckles dotted her nose. Her cheeks were rosy from the cold, and her hazel eyes sparkled. She was beaming. It wasn't a fake smile like the one the girl downstairs had. Her smile was so big you could see her teeth shining in the sunlight. I wondered what Ian was feeling. I'm sure he had seen the picture a hundred times. But I felt wobbly-kneed. *Damn, this is what love is supposed to feel like.* I've never felt that way before, not even with Sanya.

I cleared my throat and swallowed. "Did you say this is Dr. Whitely's daughter?"

"Sophie's an artist here in town," Ian said. He was so close I could feel his hot breath on the back of my neck. "I recently found one of her paintings in a small gallery downtown. You may have noticed the painting of five mangos behind my desk. That's one of hers."

"Do you know her?" My hands tingled, and my head spun. *Yup. Definitely in love.* You can't ignore all these physical signs.

"In a manner of speaking, yes, I do," he replied.

A white-haired man entered the room. Ian stepped aside, leaving me with the photo of Morey Whitely's daughter in my clammy hands.

I tried to put it back discreetly, but it fell face down when I set it on the desk. We both turned to greet Morey.

"This is Narin Roy." Ian nodded toward me. "I'll leave you two alone to talk." He left, shutting the door behind him.

Morey motioned for me to sit, so I did. With a grunt, he adjusted his daughter's picture. Morey wore a wrinkled shirt, a loose tie, chinos, and worn-out tennis shoes. His head was topped with thick, white, unkempt hair that formed bushy sideburns. His nose was flat, red, and cratered. He began to pace behind his desk.

He spoke with a thick Australian accent. "You worked with Cohen in Chicago." He didn't wait for my response. "I've talked to him before. He seemed to be a bit of a dullard, in my opinion."

I nodded and shifted in my seat.

"Cohen believes that the effects of DMTA are mediated by the Type I and Type II receptors," he said. "Although this is a hot new theory, the evidence is still not convincing that these receptors are involved in DMTA's effects on humans."

"I think the data is quite compelling, at least in animals," I replied coolly. "However, it will take time to demonstrate that the binding of these receptors causes the effects observed in Red Sky users. There is no human data."

Morey placed his hands on his desk and leaned forward. "How does binding a small pool of receptors in the thalamus produce all this?" he asked, throwing his hands in the air.

"How does it cause hallucinations?"

"How does it make you feel like a goddamn mystic?" he shouted. "How does it make you feel like you are one with the whole fucking universe?" His nose had turned a shade redder.

"I've got some theories," I replied.

Morey scoffed. "You mean Cohen and Cranston have some theories?"

I shrugged.

"Unfortunately, I've heard that the NIH didn't like your theories."

"An RFA is coming out soon," I said. "We'll see." I was surprised by how calm I felt. The urge to pull at my hair hadn't even crossed my mind during the entire interrogation.

"I've read about your theories, or at least Cranston's and Cohen's unsubstantiated speculation, that DMTA impairs sensory filtering, which causes sensory overload. In my opinion, the whole field is going to hell. This may all turn out to be bullshit in the end."

"I think that the data speaks for itself," I replied. "But you're right. There are still many gaps in the theory and numerous unknowns. Substantiation requires time and resources, and as the saying goes, the devil is in the details."

"Bloody oath!" Morey plopped into his squeaky chair and spun around. "Indeed, the devil himself resides in the bloody details." He pulled out two shot glasses, opened a bottle of Jack Daniel's Black Label, and poured out two generous shots. I reached for my glass, but Morey stopped me.

"I'm not convinced about the Cranston-Cohen model, but for now, we'll assume that it's correct," he said. I waited for more.

He continued, "We've acquired a compound that binds to the Type I receptor and boosts the affinity for DMTA by roughly 10 times. We've named it H138."

I needed clarification on this. "Like a modulator? I was unaware of any modulators for DMTA binding at either receptor."

"Now you know," he said flatly. "As far as we can tell, it doesn't affect DMTA binding to the Type II receptor. But to be bloody honest, the Type II receptor has been a pain in our arse for a year now."

I nodded.

"Our pharma lab is close to developing a compound that binds to the same modulator site, the H138 site. However, it has the opposite effect, decreasing DMTA's affinity for the Type I receptor. The goal is to mimic microdosing in a controlled manner."

He paused briefly to let that sink in before continuing. "We might be able to develop a DMTA formulation that provides the positive effects on mood and cognition without all the hallucinations and addiction." He looked at me with bloodshot eyes, trying to gauge if I understood the implications. I definitely did.

"But it would be useless if the Type II receptor is also activated," I said. "You'd never get a new drug approved if even a small percent of users end up filleting themselves."

"That's what the Cranston-Cohen model says," he replied. "We need a reliable Type II receptor blocker for this to be a feasible pharmaceutical option."

I laughed. "The world needs that."

He didn't laugh or smile. Instead, he kept staring at me. "Do you know Francis Eugene?"

That was a question I wasn't expecting. "Yeah," I muttered. "I've spoken to him a few times. He runs the program at UCSF. He's the one who discovered the two DMTA receptors."

"Do you know someone named Jason Chen?" he asked pointedly.

I didn't reply, but I was beginning to understand. Jason knew what I was doing with the Type II receptor blocker in Cohen's lab. I'm not sure if he realized how far I had gotten, but he knew I was working on it and making good progress. He probably told Dr. Eugene, who then told Morey. I wondered who else might know.

"Do you have a Type II receptor blocker?" he asked. He stared at me so hard he didn't even blink.

What could I say? I just gave a slight nod.

"Are you willing to share?"

My heart was pounding, and my hands were sweaty, so I rubbed them on my pants. I wanted to grab a handful of my hair and tear it out. I'd worn my only suit again, and now I was sweating so much I'd have to take it back to the dry cleaners.

"If you'd like to sell it to us outright, we can arrange that as well," he said. His bloodshot eyes pierced me like daggers.

"Then I wouldn't get any credit for the discovery."

"That's true," he said, shrugging. "How could we give you credit? You would have to admit that you developed it in Cohen's lab. However, if you join us, we will make sure you get the recognition you deserve."

You might ask yourself, 'Why didn't I just leave?' There were plenty of red flags. Ian was a creep, and according to my dad, he had a sinister backstory. Maru was an asshole who hated me for no reason. Morey was starting to seem like some Aussie gangster. What about that haunted clock down the hall or that evil, hungry-looking bird? That wasn't even considering that Harvester was supposedly pillaging the Amazon rainforest and screwing the indigenous tribes of Amazonia.

"One hundred and eighty thousand per annum," Morey said, breaking the silence. "There's HR bullshit, but that's what Ian says we can do. Do you accept?"

I didn't say anything. I looked over and saw Sophie smiling at me. She was smiling at me! If I could make something of myself, become famous, and save the world, I could have a girl like her. What was I saying? I deserved a girl like her. This was my only shot. It was either this or a knife in the chest. Those were my only two choices.

I reached out and picked up the shot glass.

"You accept this offer, and you're my boy. You're not Ian's boy. You belong to me. You answer only to me. Is that clear?"

I nodded.

"On day one, I expect to have that Type II receptor blocker," he said.

Our glasses clinked together with a pleasant sound.

We both threw our heads back and swallowed.

Part II

Eleven

Now that I've been writing for a while, I do have a few complaints. It's distractingly cold in this interrogation room. I spent last night in a cell wearing a thin orange jumpsuit, and I'm still in it. I've been rubbing my hands together to stay warm. Also, this uncomfortable chair is the only one in the room, and it rocks back and forth on uneven legs, which drives me crazy. I imagine they made this room uninviting to get suspects to confess more quickly. It's a form of mild torture: harsh, flickering lights; cold, dry air; and a hard, uneven chair. Well, I'm sure they know what they're doing.

Another annoyance is that this room smells of bleach and lavender cleaning products, with a faint scent of tobacco. Cigarette smoke is one of my biggest migraine triggers. My left temple is already beginning to throb. And here I am without my migraine rescue medications. Would they give me something to take if I needed it?

I can't help but keep staring at the dark window occupying one wall. I picture a bored officer on the other side drinking lukewarm coffee and trying to stay awake. They have me on suicide watch, which—considering the situation—seems like a good idea. They probably think I'm upset about yesterday's tragic events. But I'm not. I'm just completely numb. That's the power of Red Sky. When Red Sky is in control, you can't feel a negative emotion, even if you want to.

Officer Johnson came in a few minutes ago. I hadn't met him before. He looks like a kid straight out of high school, peach fuzz and all. He brought me a sandwich on a paper plate and a Styrofoam cup with water. The sandwich has two slices of white bread, a sad-looking leaf of lettuce, and what appears to be a slice of turkey. I haven't eaten it yet, and I doubt I will.

He asked if I needed anything. A restroom break seemed like a good idea, so I went. I also wanted to call my parents. I could tell he wasn't expecting that. After fumbling, he said he'd need to ask Officer Coffin if that was okay. That was ten minutes ago, and he hasn't come back. So, I guess I'll keep writing.

A few things about what just happened might confuse you. They confused me too after I finished with Morey. I called him Morey when I was at Harvester, so it feels strange to call him Dr. Whitely here.

First, if Jason Chen knew I had a Type II receptor blocker, why didn't he take the opportunity to bring me to San Francisco to work with him? He mentioned a hiring freeze, but that seemed like a stupid reason not to hire me, especially since I'd made a major discovery. I quickly realized that Jason was starting his academic career and didn't want me to overshadow him. He had always been jealous of me. That "Rah-Rah" and "You can do it, Narin" was all an act. He was hoping I'd fail and disappear. The last thing he wanted was competition.

Second, why had I fallen head over heels for Sophie? You'd think I'd never seen a picture of a beautiful girl before, but you'd be mistaken. There was a time in my life when I became very interested in pornography and would spend hours a day looking at it. This was during the early days of dial-up internet, so a couple of hours meant maybe a dozen nude pictures that loaded one line at a time. When broadband arrived and allowed videos to play smoothly, I realized something about myself: I hate seeing bodily fluids in any form. That's why I stopped watching porn. You never know what disgusting thing might pop up.

Getting back to Sophie, there was something special about her. She wasn't like Sanya. I wasn't sure what it was, but after that first glance, I knew I was in love with her, as crazy as that sounds.

After leaving Morey's office, I went to find Ian. I couldn't remember which office was his, but luckily, his door was open. I peeked in, but he wasn't there. After a moment of hesitation, I stepped inside. The bird watched me as I crossed the room, but it didn't make a sound. I walked behind Ian's desk to look at Sophie's painting of five ripe mangoes stacked together. I don't know much about art, but it looked well-painted. The mangoes resembled mangoes, which must take skill.

I ran my finger over her signature. 'Sophia Whitely.' I stood there, burning each letter, each mark, each curve into my memory. As I examined her signature, the gears in my mind started turning. How was I going to win her over? What would it take? Hell, I hadn't even met her yet! How would I arrange a meeting? Also, Ian seemed interested in her. I'd have to step up my game if Ian were my competitor. Once she was mine, what would I tell my parents? Not only was she white, but she was an artist. That wasn't going to go over well with them.

I was jolted out of my reverie when that damn bird squawked. I decided to leave, wanting to avoid seeing Ian again. I needed to think things over. As the elevator doors closed, I heard the grandfather clock strike noon.

As I drove home, my mind was a mess. I'd just landed a new job promising good pay, professional recognition, and—most importantly—a chance to develop a drug that could change the world. So, what was the problem? I should have been excited, so why did I feel like I had a noose around my neck?

When I got home, I parked on the street. A car was in the driveway, leaving no space for another vehicle. I worried that the Porsche might be stolen or damaged, but I didn't have many options. As soon as I entered the house, I began to salivate. Whatever was cooking smelled amazing. I found my housemate comfortably sprawled on one of the couches in the living room, reading a book. He jumped to his feet when I came in and greeted me.

"You must be Narin Roy. Ms. Flowers told me to expect you." His voice was soft and silky, not the stern one I'd anticipated from a man his size. I shook his outstretched hand. "I am Marcus Sanquel Arroyo. Please call me Sanquel. I will be your housemate."

How do I describe Sanquel? He could easily play the role of a noble king in a movie. He wore his long black hair pulled back. He had a well-defined face, a trimmed mustache, and a set of shiny, perfect teeth. He had what Mom would call a 'tapered figure.' His shirt was tight around his chest.

"I am very pleased to have a housemate," he said. "It can be quite lonely here with no one to talk to." He gave me a broad smile. "I understand that you are relocating from Chicago."

I nodded. Sanquel's pronunciation of Chicago had a melodic quality.

"I have many close friends who live in Chicago, mostly starving artists I met through my father." Sanquel laughed. "Expats like me enjoy staying in touch." An alarm beeped in the kitchen. "You must excuse me for a moment." He rushed to the kitchen and called out, "I've been preparing a meal, nothing too extravagant, but I always make far more than I can eat. I was hoping you would join me."

"It smells wonderful," I said, realizing I was starving. I sat on the worn green couch, which turned out to be surprisingly comfortable.

I heard dishes clattering in the other room. Should I offer to help? I was about to ask when Sanquel reentered the living room, wiping his hands on a towel.

"How long have you lived here?" I asked.

"Many months now, though my work keeps me away from my bed many nights," he said and sat down next to me. "I am a slave to the company that employs me. That is, of course, my fault and my fault alone. I am a part owner of this company, so you could say that I am both a slave and the enslaver." He laughed and continued, "My partner and I relocated to St. Louis when we received seed money to develop a factory here. We make robots." To illustrate his point, he moved his arms and legs mechanically and laughed again. "I work primarily on the software side of things, while my partner is a brilliant designer. We each do our part and stay out of the other's way. But honestly, we work very well together."

"That sounds interesting," I said. "That's an area about which I know virtually nothing."

"But you're a scientist, too?"

I wasn't expecting him to know that. "How did you know?" I asked.

Sanquel tapped his head. "I have great intuition about such things."

I hesitated for a moment. "I'm studying DMTA, I mean, Red Sky. DMTA is the scientific name. I work with animal models, primarily mice."

"Seeing those mice get high must be fascinating." He laughed again. "Are you working at the colossal medical center down the street?"

Here it comes. He will eventually find out, so I might as well tell him. "I'm starting at Harvester Pharmaceuticals." I tried to gauge his reaction. If he was appalled, he didn't show it. "I'll be starting next week."

"How about we keep this conversation going with some food and drinks?" Sanquel stood up and gestured toward the kitchen.

A delicious spread was laid out on the kitchen table. I sat down.

"We start with pozole rojo," Sanquel explained, "It's a traditional Mexican stew. I imagine it won't be too spicy for you. I hope I'm not being presumptuous, but I suspect spicy food doesn't bother you." He scooped out two generous bowls as he spoke, setting one before each of us. I blew on my stew, and steam rose from the surface. I'm not a big fan of spicy food, but I didn't mention it to him.

"My father was a very talented man," Sanquel continued. "He loved to cook and taught me from a very young age. Men of his generation rarely learned to cook in Mexico; I am fortunate for his total disregard for convention."

I nodded as I carefully tasted the soup. It was delicious.

Sanquel pulled up a chair and sat next to me. "By profession, my father was a painter of some renown. Unfortunately, he was a man with strong opinions. He was actively involved in the student, indigenous rights, and global Communist movements. He supported any ideology that antagonized the establishment. But unlike others with that political persuasion, he lacked the sense to restrain his words. He was foolish that way. Luckily, I have learned from his mistakes."

"Did you grow up in Mexico?" I asked.

"As a result of my father's incautious nature, we were forced to flee Mexico under significant duress," he said. "I was seven years old and quite shy. He insisted that I go with him into exile in Spain. We lived in Valencia. My mother and sister stayed behind in Mexico. He needed someone to assist him and chose me because of my pliability. He knew he could control me. My mother refused to come with us. She is a devout Catholic who believes the Spaniards have lost their way. She thought Spain would be a corrupting influence. For some reason, her concerns did not extend to me. My father and mother had drifted apart emotionally long before, so it was no surprise that they would drift apart physically as well."

"Do you still have family in Mexico?" I felt my mouth burning, so I took a big gulp of water.

"My sister is currently in the States. My mother lives in Mexico City, in the same house she shared with my father. Even there, she laments the materialism and lack of faith. I believe she would be happiest in a rural village, where the influence of church and family remains strong."

"I was born in India," I said between swallows. "I know how hard it is to come to a new country."

"How old were you when you came over?"

"I was three years old and didn't speak English," I said. "I learned it by watching television. It took me some time, but I learned how to fake it till you make it."

Sanquel laughed. "It is hard to be uprooted like that. Of course, at age three, it might not be so bad. For me, it was a tough transition. My father, may he rest in peace, had many talents, but he wasn't cut out for raising children. Providing comfort and affection were not in his nature. When the local kids teased me for my strange accent, my father told me to act like a man. I was seven years old and naturally scared. For years, I lived a lonely life. I worked in my father's workshop, cleaning his brushes and preparing the paints. I cataloged all his artwork. I even managed his finances since he had no head for that. I still remember the fumes from his studio. They made me sick every single day of my childhood. That's what finally did him in—those toxins from inhaling in a cramped, unventilated studio for so many years."

Sanquel paused for a moment. "But over time, I adapted, Narin. I developed a knack for—what is the phrase you used? I learned to 'fake it till you make it.' Yes, that describes me perfectly. Soon, I was speaking like a native. I jogged to clear my lungs, initially for short distances, then for hours at a time. I became quite the athlete, something no one would have guessed could happen if they had known me as a young child."

I scraped the bowl with my spoon, chasing the last bits of the stew. "This is excellent," I exclaimed. "I've never had anything like this before. Tell me about Valencia. I've never been to Spain."

"It's a beautiful city," he said with a smile. "Spain has many stunning locations, but Valencia stands out as unique. The Romans founded it centuries ago, and it remains a hub for shipping and trade. It's a very diverse and lively place. I still miss being there."

"It sounds very different from St. Louis."

Sanquel cleared the bowls and brought over a plate of tamales. We each took one and started eating. "St. Louis is not so bad," Sanquel sighed. "Over the years, I've lived in much worse places. Everyone complains about the humidity and mosquitoes, but those things don't bother me. I dislike driving here. I've taken up biking everywhere to avoid dealing with the terrible drivers. Outsiders often complain about driving in Chicago, but at least in Chicago, people tend to pay attention to what they're doing. They don't make U-turns in the middle of the road in Chicago, or they'd get squashed like a bug!" He slapped his hands together for emphasis and laughed. "Some of the major roads here even change names without warning. They must assume that only locals drive these roads. And another thing," he said, waving his utensils as he spoke. "I hate how streets are blocked off with these wrought-iron fences or rusty chains. They go to the trouble of building a nice road, only to block it off. It's quite dangerous riding at night when a chain appears stretched out in front of you while flying at full speed. I guess they do that to keep out the undesirables. I wonder sometimes if I'm one of those undesirables. The strangest thing to me is that those most inconvenienced live in neighborhoods that are blocked off. They are the ones who must circumnavigate these barriers daily. It is quite—what is the English word—" He struck his forehead. "Provincial. The attitude of the people here is quite provincial."

I finished my tamale and reached for another, realizing I hadn't eaten this much in months.

Sanquel continued, "Frankly, I haven't spent much time trying to meet new people. The only person I've become friendly with outside of work is Ms. Flowers. She seems just as lonely as I am."

"Thank you so much for this wonderful meal," I said. "I feel like I'm going to sleep well tonight."

Sanquel laughed again. I helped him clean up and put away the plates and utensils.

Twelve

I woke up early the next day and lay in bed, mesmerized by the dust motes visible in the sunlight filtering through the blinds. The tiny particles seemed to bounce around as if they had a mind of their own. I thought about my grandmother's beliefs in invisible spirits that control our destiny. I wondered if I was just one of the countless specks of humanity being tossed about by forces beyond my understanding.

My phone rang, and I didn't recognize the number. I might have ignored it, but I was worried Harvester could be calling, so I answered. It was Ian.

"Morey says that you've accepted our offer," he stated, his enthusiasm evident.

"I did," I said. "I'm excited about the idea of turning DMTA into a pharmaceutical. I want to make sure you all head in the right direction." I was about to say more, but Ian stopped me.

"I'm just the legal grunt. You will need to take that up with Morey," Ian replied. "I wanted to see if you'd be interested in a friendly poker game tonight with some of your future colleagues at Harvester. It's our Friday night ritual."

I enjoyed playing poker and thought I was good at it, so I accepted the offer. He gave me the time and address.

I was just about to hang up when Ian said, "Hey, I have a favor to ask you."

"What's that?" I asked.

"Do you remember Professor Goldblum? Your father's old friend?"

"What about him?" I wondered, confused. Where was this headed?

Ian replied, "You promised to visit him when you came to St. Louis."

This set off some alarm bells. Had Ian been spying on me? And if so, why?

"What's that to you?" I answered testily. "I won't waste time visiting one of my dad's old friends." I didn't mean to get irritated and kicked myself. The last thing I wanted was to sound like an asshole to a guy I barely knew and who had just gotten me a job. I heard my sister's voice in my head: 'Try to act more normal.' *Good advice.*

Ian paused for a moment, as if weighing his options. Then he said, "I think you should go see him tonight at six o'clock. It won't take very long. Just give him a call. You may be surprised at what you find," he said mysteriously. "Trust me."

I was skeptical but also intrigued. "Sure, Ian," I said. "I'd be happy to." There was no need to get annoyed with him. Visiting Goldblum wouldn't be too much trouble, and my dad would be happy if I did it.

After hanging up with Ian, I found the phone number my dad had given me. I wasn't exactly sure what to call him, Mr. Goldblum or Dr. Goldblum. But my dad had introduced him as 'Professor Goldblum,' so I went with that. I called him and asked if I could come over at six that evening. He sounded excited and gave me his address in South St. Louis.

I headed to the kitchen. Sanquel was out. I poured myself a bowl of cereal and checked my emails on my phone. Mom and Dad had arrived safely in Kolkata. They had spent the last few days visiting Mom's father, who, after breaking his hip last year, had been languishing. But she said he was slowly recovering and walking with a cane. He had even resumed some advocacy work for the poor and disadvantaged.

She mentioned visiting relatives and overeating. She described how much Kolkata has changed. She and Dad were sending emails from a newly opened internet café. A couple of things were noticeably missing. There was nothing about Dad's family, the ceremony with the ashes, or all the legal issues Dad had gone there to handle. She also didn't say when they'd be back. I responded briefly, telling her I had found a job in St. Louis and would stay here for the foreseeable future. I didn't give any specifics. There would be time for that later.

When evening came, I slipped into an old, ragged hoodie and sweatpants and headed out. The professor's house was a simple red-brick ranch-style home. A weathered fence encircled the yard. Several well-tended potted plants sat on the porch. A wooden swing hung from rusty chains. I pressed the cracked doorbell, curious about what "surprise" might be waiting for me.

The professor, wearing a wool sweater vest and corduroy pants, opened the door.

"So good that you could come. Please come in," he said. "There is someone I would like you to meet."

The fading evening light streamed across the old floorboards of the living room. Books lined the shelves and tables. Only two chairs were empty, and Sophie sat in the one by the window.

"May I introduce Ms. Sophia Whitely?" the professor asked.

"Uh, hello, Ms. Whitely," I muttered incoherently.

However, my internal monologue was crystal clear: *You little piece of shit, Ian! What the hell? You set me up!* This wasn't how I imagined my first encounter with Sophie. I looked like a total slob, and I hadn't prepared anything funny or clever to say.

"It's Sophie. Please call me Sophie." Her voice was gentle and a bit husky.

"Please sit next to my friend Sophia," the professor said. "I'm making tea. It'll be ready shortly. I've been looking forward to your appraisal." The professor left for the kitchen.

I sat next to Sophie, even though I didn't have much choice; only two chairs weren't covered with books. I forced myself to look at her. Her hair looked different from how it appeared in Morey's photo. Her thick, dark hair was now shoulder-length with bangs in front. The sunlight from the window highlighted the reddish streaks in her hair. I've always liked long hair, but Sophie was still stunning just the way she was.

I struggled to find the right words. "Did you say your name was Sophie?" I asked.

She nodded, cupped her hand to her mouth, and leaned in as if sharing a secret. "Everyone calls me Sophie, except for him, for some reason." She nodded toward the kitchen, let out a small laugh, and looked out the window.

"So—how do you know Professor Goldblum?" I asked, pulling my hair so hard it hurt. I finally paused and wiped my sweaty palms on my pants.

"I live right across the street," she said, motioning with her head. "The professor and I have mobility issues, so these ugly ranch-style houses work well for us."

I looked at her expectantly, hoping she would elaborate, but she didn't. The professor was taking forever to return. I gathered the courage to look at her more closely. She wore a yellow summer dress held up by a thin strap around her neck. The contours of her legs showed through the delicate fabric. Her face and shoulders were tanned, with brown freckles around her nose and spreading across her forehead, shoulders, and back. Her cheeks were flushed and dotted with peeling skin, as if she had been out in the sun too long.

She looked at me appraisingly, her hazel eyes shining with curiosity. "You know something?" she began. She was interrupted by the sound of cups clinking in the next room. "I really should go help him," she said.

She grasped the armrests and pulled herself up. Her steps were unsteady as she moved toward the kitchen.

The professor stood at the kitchen door, holding a tray with three cups and a teapot.

"Let me help you with that." I jumped up from my chair, hurried over to assist, and nearly knocked over Sophie, who grabbed the arm of the couch to steady herself.

"I'm so sorry," I squeaked in panic. Luckily, she was still standing.

"Nonsense," the professor replied. "You two are guests. Please sit." We moved back to the chairs and sat down again.

The professor's hands trembled as he poured the tea, but he managed to fill the cups without spilling. He cleared books off the couch, adjusted his sweater, and after a moment of fumbling, placed a pair of bifocals on his nose. He sat down and took a deep breath.

Sophie and I each grabbed a cup. I blew on my tea as she took a sip.

"I'm so happy, Narin, that you had the chance to come, try my tea, and meet my good friend Sophia. She's been very thoughtful to me over the past few months, always checking in on me and sharing a cup of tea."

"And stealing a book or two," she added with a smile that made my insides turn to mush.

"I have far more than I need," the professor replied. "The problem with books, as with all possessions, is that they isolate you from what is real. I'm better off without them. I'm sure that soon enough, I will be gone, and all these books," he waved his arm in a sweeping gesture, "will scatter to the wind."

The professor tilted his head slightly and leaned forward. "So, Narin, does my choice of tea beat the almond and cinnamon that your father prefers?"

"Um—it's good, I mean, perfect," I said. I looked at Sophie to see if she thought I was a clumsy idiot. Her face stayed inscrutable.

The professor chuckled, sounding like a rusty door hinge. "I miss the days when I would talk over tea with your parents in Chicago. I still chat with your father often, but such long-distance conversations can never compare to those over a good cup of tea. But my comfort, at least for today, is that you are here with me. You remind me so much of your dear father."

"I've been told that," I said.

"Do you know how your father and I first met?"

"No," I replied. My eyes kept darting between the professor and Sophie. "Not really."

"Our history began under challenging circumstances. I faced harsh accusations. After I published *The Value of Nothingness*, my critics accused me of the deadly academic triad: pedestrianism, pedantry, and plagiarism. My chairman believed I should be forced to relinquish my tenure. Your father sat on the tenure committee and was my strongest supporter. His support gave me several good years to enjoy his company. Only after the chaos did I get to know your father and realize how much we had in common. Eventually, my chairman, who refuses to accept defeat, pushed me out of my position. Luckily, I found work here in St. Louis."

I looked at my phone resting on my lap. Ian told me the card game started at seven o'clock. The location was thirty minutes away, so I needed to leave soon. I had to find a way, an excuse, to see Sophie again.

The professor droned on. "Over the intervening years, many of our conversations have centered on you." He winked at me. "That's why, in some ways, I feel I already know you." With this last line, the old professor leaned forward and lowered his head. The shift in his manner changed the tone from that of a kind grandfather to a prickly uncle.

I wasn't sure how to respond, but luckily, Sophie spoke up. "I haven't found a copy of your book in any of the local bookstores. I'd hate to have to order it online."

"My dear," the professor said, "if you'll be patient with me just a moment, I'll find a copy for you that's hidden away on one of the shelves in my study."

He pushed himself off the couch and moved to the back of his house. I kept licking my dry lips, opening and closing my mouth. My mind was blank. Sophie stared out the window. The silence grew more unbearable by the second. Then I remembered she was about to tell me something just before the professor came in with the tea. I opened my mouth to ask what she wanted to say earlier, but then Professor Goldblum returned and handed Sophie a copy of his book.

I took this opportunity to leave. "Professor, I'm afraid I have other commitments tonight." I set down my teacup. "I wish I could stay longer."

I didn't want to cause any more damage tonight. I'd come across as stupid, boring, and a complete klutz.

The professor asked, "Will I see you again before you go back to Chicago?"

"Actually," I said, glancing at Sophie, "I'll be staying in St. Louis for the foreseeable future."

"Wonderful," the professor said. "Your interview must have gone well. You are welcome to come at any time. I would love to have you."

I hurried toward the door, and to my surprise, Sophie followed me outside. She handed me the book and carefully navigated the staircase, holding onto the railing as she descended. At the bottom, I returned the book to her. She nodded and offered a polite smile.

"How often do you visit him?" I asked as we walked toward my car.

"I bring him dinner on Sundays," she said. "He doesn't eat much, but I try to give him a good meal every few days." She looked back at the house. "I wish I could do more. His posture is getting worse, and I'm worried he might fall someday."

I seized the opportunity to see her again. "If you don't mind, I'll bring him something too."

"That would be great," she said brightly. "I'll throw together a salad with stuff from my garden. He likes to eat around eight o'clock."

We arrived at my car. The red Porsche sparkled in the fading sunlight, and a flicker of disdain crossed her face.

"You were about to say something," I said, partly to distract her. "In there."

"Was I?"

"You started—you began to say something. I think I interrupted you."

"I was just going to say that you look like my fiancé. My former fiancé, I should say. He died in a car accident."

At that moment, I should've said, 'That's so terrible. I'm so sorry for your loss.' Instead, I blurted out the first thing that came to mind. "You were—you were going to marry an Indian guy?"

"Yeah, his name was Ravi Reddy."

"You know what they say. We Indians all look alike," I said, letting out a fake laugh.

"No, no, you don't." This brought out a genuine smile for the first time. Damn, did her eyes sparkle!

"It was a pleasure meeting you, Sophia, I mean Sophie."

She laughed at this. "I guess he's decided that he prefers Sophia. Everyone, even my parents, calls me Sophie."

"Well then, it was a pleasure meeting you, Sophie," I said with a smile.

We shook hands, but it wasn't a typical business handshake. She held onto my hand longer than necessary, and my whole body flushed with embarrassment. If I'd been braver, I might have gone for a hug. Maybe I could have kissed her on the cheek, like Europeans do. But no, the handshake was more than enough for now. Any more than that, and my heart would have jumped out of my chest.

I waited in my car until she went back inside her house. If I waited any longer, I might come off as a stalker. Plus, I had a poker game to get to.

Thirteen

A man about my age with an athletic build, coarse black hair, and a Neanderthal-like face stood outside the apartment building, smoking a cigarette. He wore an untucked flannel shirt, ripped blue jeans, and sandals.

"Are you here for the game?" he asked as I approached.

I nodded.

He dropped the cigarette and crushed it under his heel. I followed him into the apartment building, which looked dull and shabby. I could hear Ian's voice above the others, like a barking dog, even before we went inside. I was still furious with Ian for setting up that meeting with Sophie. What game was he playing? Was he trying to help me or screw me over?

The apartment had a strong male smell, like dirty socks and nacho cheese. Luckily, no one was smoking. A bright overhead light lit up the table where the players sat. Ian looked up at me expectantly when I entered, and I glared back at him.

"Ah, Floyd, you found my friend, Narin Roy," Ian said. "Please have a seat."

Floyd sat down, and I took the seat across from Ian at the table. Black-and-white chips were stacked in columns to Ian's right, while three battered decks of cards lay to his left. He wore a dealer's cap crooked on his head, casting a greenish shadow over half his face.

"Boys, this is my friend Narin Roy. He has decided to join Harvester, so I thought this would be a good way for him to meet some of the characters in our motley gang. Let's all make him feel welcome."

The others didn't even look up; they were busy counting their chips with grim expressions. From the muffled conversations, I learned that the white chips were worth $5, and the black chips were worth $10. I hadn't expected us to be playing for such high stakes. I exchanged some money and received a stack of chips totaling $100.

"Introductions," Ian mumbled as he shuffled the cards. "To my far left, and your immediate right, is Chris Vender." Ian paused. "He comes to us via beautiful Lausanne, Switzerland. I don't exactly know why he's here."

"Because you like to take my money," Chris said in a generic European accent. He had short, flaxen hair, a narrow face, and round spectacles. The sleeves of his brown dress shirt were rolled up, and his tie hung loosely around his neck.

"And to my left is Carlos DeAngelo." Carlos nodded in acknowledgment. He also wore a dress shirt and tie, suggesting he had just come from work. Carlos had dark skin and soft brown eyes, along with a laid-back demeanor.

"Chris and Carlos run the Pharma lab," Ian added. "They produce the materials that Morey and the other division heads order. You should stay on their good side, Narin, because they control which projects get priority."

"Yes, Narin," Chris said in an unfriendly tone, "you should stay on our good side."

Carlos shook his head. "We're just short-order cooks. We're here to provide whatever you need to complete your work."

Ian took a deep breath. "To my immediate right is Alex Lifer, an old friend. To our delight, he has decided to join us for the next few weeks."

Alex leaned back in his chair. He was tall and thin, with curly blonde hair and beads hanging around his neck. His clothes looked worn but somehow stylish.

Chris interjected, "You didn't mention the most fascinating thing about Alex. He's the only one here who has experienced the pleasure and privilege of dating Morey's daughter."

"Make that pleasure and pain," Alex muttered under his breath.

Ian looked at me with concern.

Chris pressed on. "Or I should say, the only person I know who has dated Morey's daughter and didn't die a horrible death. Now, remind us, Alex, what ended that lovely, sordid affair of yours?"

"Let's not air our dirty laundry here," Ian interjected. "That was a long time ago."

Alex shook his head. "I found out she was as crazy as her father."

Ian glanced at me again before turning away. He was obviously trying to manage the situation. He hadn't expected them to bring up Sophie's past at the poker game.

"And finally, we have Floyd Fox," Ian quickly added. "He is the owner of this fine establishment."

Floyd quietly twirled the chips in front of him.

Ian shuffled once more. "We play Texas Hold'em." Ian looked at me again and asked directly, "Are you familiar with the game?"

"I've played before," I replied, surprised by the high pitch of my voice.

"Hand rankings are as follows: royal flush, straight flush, quads, full house, flush, straight, three of a kind, two pairs, and high card. The minimum bet is five dollars. There are etiquette points you'll learn as we go along."

We mostly played in silence. Ian mentioned that the Cardinals were two games ahead of the Reds and in first place. He asked Carlos how his Dodgers were doing, but Carlos just laughed.

"Mark my words," Ian said. "2006 is the year. They're going to win it all. It'll be the first time since 1982, and we'll all be here to see it happen." No one challenged Ian's bold assertion.

Chris added, "I doubt that, aside from Carlos, anyone cares, but the World Cup will be incredible this year. Many teams could come away with the trophy."

The conversation shifted to work gossip (who was dating whom), politics, and the ongoing war in Iraq, but their words lacked conviction. Only the game—the ever-changing cards and pot—held everyone's attention—everyone except me. I struggled to stop thinking about what Alex had said about Sophie.

To clear my mind, I focused on the other players, paying close attention to Alex. He reminded me of people I'd disliked my whole life. Alex was the kind of guy every girl would find attractive. But the more I watched him, the more I realized I knew something about him that he himself didn't recognize.

Alex had a clear tell. Whenever he had a winning hand, he would lean back in his chair and shake out his shaggy hair. This secret sign showed that I could beat this new rival if it came to that. Keep your eye on your enemy, the Desert Prince had instructed me. *I'm watching you, Alex.*

I didn't notice any other clear tells. Floyd only smiled briefly after winning. Chris and Carlos furrowed their brows as they calculated the odds with each flip. Ian played a sloppy, reckless game, challenging the others with outrageous bets and eventually losing all his chips as the evening went on.

The game was over. The only clock in the room blinked 12:00, the same time it showed when we started. Chips were sorted, and money exchanged hands. I'd broken even.

Ian walked out with me. "Did you enjoy the game?" he asked playfully.

I waited until we were out of earshot of the others. "What the hell, Ian?" I was furious.

Ian took a step back. "What?"

"Why didn't you give me a heads-up about Sophie?"

He smiled. "I thought you'd like to meet her," he said.

I threw up my hands. "Like this?" I gestured at my ragged clothes. He grabbed my arm and pulled me close without any regard for boundaries.

"Sophie's been lonely since she lost her dearly departed fiancé," Ian crooned. "She had a falling out with Morey, and they no longer speak to each other. So, keep going over there every Friday evening. You don't have to say much. Just be there to listen." He let me go and paused for a moment. "You don't have to impress her yet, Narin. Just be her friend. Nature will take its course."

What the hell? Why was he trying to bring us together? I was about to ask him all of this when a honk from Carlos' passing car interrupted us.

"I'm sorry about Alex," Ian said. "I didn't realize he would be here today. I wouldn't worry about anything he said."

Ian turned and walked away.

"Hey, Ian," I yelled. "Do you mind if I hold onto the Porsche a little longer? I'll go car shopping after I get my first paycheck."

"No problem," Ian said. "I have three of them. You're driving the base model. This," he said, running his hand over the top of his sapphire-blue car, "this sweetheart is the premium. I'd highly recommend it for you someday." He got in, revved the engine, and peeled away. It struck me as odd that Porsche still had a model called 911 after what happened in 2001. But they must be doing something right. It sounded like Ian was a loyal customer.

I sat with my hands on the steering wheel and made a decision. Instead of heading home, I turned back toward South City. When I reached Sophie's house, I slowly drove past it. All the lights were out. The professor's house was dark, except for a flickering porch light. I parked a block away and walked back to Sophie's house. The area was quiet. I wandered around and looked into each window. The blinds were all closed, and all I saw was my sad reflection against the dark glass.

Panic overtook me, and I snapped back to reality. What was I doing? How could I explain myself if she saw me? That would instantly ruin my chances with her. Imagine the humiliation! I could even get arrested. I felt sick to my stomach.

I drove home as quickly as I could.

I climbed the stairs, entered my room, and went to bed. The hoodie was uncomfortable and bunched up behind my head, so I took it off. Soon after, I found myself on a tall cliff overlooking the desert.

The Desert Prince looked down at me, his face grim. He said nothing.

I broke the silence. "I've been practicing," I said.

He responded sharply, saying, "You've also been wasting your time on frivolity."

Who was he to tell me how to spend my time? "What? The poker game? Am I not allowed to have fun?"

He commanded, "Show me what you have learned!"

With that, I found my knife in my hand and took my stance. He shook his head, and I wasn't sure if that was a sign of approval or not. His face remained impassive. Without speaking, he began to circle me. I was still upset that he had scolded me.

He gently pushed my arm down. "It is best to conceal the knife if your opponent does not display his weapon. There is no need to give him any advantage. Drawing your knife will distract him, making this an ideal time to attack. If you miss this opportunity or if it is unavailable, wait patiently for him to launch an attack. That is another moment when he will be vulnerable to distraction. To strike effectively, he will have to break his stance. Then, launch a quick counterstrike before he regains his footing."

I nodded.

"There's a gap between the decision and the action. You must close that gap. Do you understand?"

"I understand, Sir!" I shouted. "Translate my decisions into action!" I loved how that sounded.

He continued, "When the aggressor approaches, make a quick judgment about your opponent. Look into his eyes to gauge his emotions: aggression, hatred, uncertainty, or fear. This is one of the most important decisions you will make. Keep eye contact, but also watch his hands. Notice how and where they are moving. Often, the right hand will be near a hidden weapon. Check the outline of his clothing to see where he might be hiding something lethal."

Everything he said seemed really important, and not just for knife fights.

"Always keep moving. To stay still is to let your opponent strike. Move forward and backward, and circle left and right. But always maintain your balanced stance." He demonstrated the movement. "Stay light and agile. You don't need to plant your feet firmly to attack your opponent, as you do when fighting with your fists."

I should say that I have never hit anyone in my life.

He continued, "Observe your opponent to identify which move gives you the greatest advantage. If your opponent has a knife, keep your distance. If your goal is to kill your opponent, target their neck or torso, but the most effective tactic is to aim for the arm holding the weapon. This is called 'defanging the snake.'"

"Defanging the snake," I repeated. "I like that." And I did. The sound of it sent a thrill through me.

He stepped forward and towered over me. "The world is wicked, full of treacherous and dishonorable men. Prepare now, strengthen your body, discipline your mind, and stay constantly vigilant, for your enemies are drawing ever closer."

I looked at him blankly.

"You will find what you need in the substructure where you reside."

He meant 'basement,' but I wouldn't correct him while standing in his shadow.

I woke up multiple times that night. I tried to recall everything the Desert Prince had told me, but my mind kept drifting back to Alex and Sophie. I replayed every word Alex and Chris said during the poker game, eager to learn more about whatever 'sordid affair' they were talking about.

I woke up late on Saturday, and Sanquel was reading the newspaper in his usual spot. "How was your poker game?" he asked.

"Poker is an interesting game, very psychological. I would have won more if I knew them better." I was still thinking about Alex and Sophie.

"Have you ever been married, Sanquel?" I asked.

"No, my friend," Sanquel smiled. "I don't think marriage is in the cards for me."

Fourteen

I drove to Left Bank Books and asked the person at the desk where I could find *The Value of Nothingness*. They had only one copy left, a used book that was pretty worn out. Almost every page was highlighted, and entire sections were underlined.

At dusk, I collected my dirty clothes and headed down to the basement. A set of weights sat in the corner. This must have been what the Desert Prince was talking about. While the washer and dryer ran, I worked out. I was glad no one was watching me. I felt incredibly weak. My muscles were in full revolt by the time I finished.

On Sunday morning, I went into the basement and returned an hour later, covered in sweat.

I hung out in my room for the rest of the afternoon. I received an email from my mom asking about my new job. Dad had warned me about Ian, so I didn't want to tell them I'd accepted the job at Harvester, at least not yet. Mom also mentioned that Deepa had met a nice boy named Rajeev, whose father is a cardiologist in Los Angeles. I was surprised Deepa had told them. Things must be getting serious. I called my sis, but she didn't answer.

I lay in my bed reading the professor's book. His ideas now seem cliché because everyone is into mindful meditation. When it was published, it captured the 1960s zeitgeist, or so I'd heard. People appreciated his view: that one could find peace without relying on anything outside oneself.

They just needed to focus on the empty spaces (nothingness) around them to achieve stillness of mind.

I skipped parts that seemed boring. He was trying too hard to prove that each religious tradition taught the same thing. In my opinion, religion often causes division and fanaticism, fueling hatred and conflict. So much for nothingness.

I fell asleep reading, and when I woke up, the sky was changing from reddish amber to deep indigo. *Shit. I'm going to be late.* I quickly put on slacks and a dress shirt, grabbed the flan Sanquel had made for me, and headed to the professor's house. Sophie opened the door for me. She was wearing a black tank top and denim shorts, and her face was flushed.

"Hey," she said brightly. "Come on in. Whatcha got there?"

I gave her the dish. "My housemate made flan."

"Great!" She took it into the kitchen, and I followed her. The professor was examining a bottle of wine. When he saw me, he came over and shook my hand.

"So kind of you to come over, my boy. If you don't mind, maybe you could help Sophia bring these dishes out to the dining table."

"Sure," I replied.

I helped Sophie set the table while secretly watching her every move. Even with a noticeable limp, she moved with impressive agility.

"I'm so glad to be out of the house," Sophie declared. "I've been painting all day. I absolutely hate painting. It requires so much patience and precision." She shuddered.

"Why do you do it then?" I asked.

"Because it pays the bills," she said. "And because I'm a stubborn bitch," she whispered to me so that the professor, who was still in the kitchen, couldn't hear. "My teacher told me that I didn't have a knack for it. After that, I was determined to prove him wrong. I kept refining my technique until I got good, really good, knock-it-out-of-the-park good. I'm pig-headed that way."

She barely said two words to me the other day. Why the sudden change? Or maybe she was friendly that day, and I was so embarrassed and tongue-tied that I misunderstood the whole situation.

The professor entered the dining room, and we sat down for a meal that included beet salad, a cheesy pasta dish, and asparagus, all served with red wine. The dessert was the flan I'd brought.

"Did you grow up here in St. Louis?" I asked Sophie.

She poured herself a glass of wine and gulped it down. "I was born in Sydney, Australia. My mum is from the States, and she was homesick, so we moved back here when I was twelve. It wasn't too bad. America has a lot to offer. I still did everything I liked: biking, hiking, and lots of sports, anything that kept me outside. The only thing I missed was the surfing."

"I don't detect an accent," I said.

"It comes out once in a while," Sophie said, "especially when I've got myself a gutful of piss." She suppressed her laughter. From her look, she was already headed in that direction. I wondered how much alcohol she'd had before I arrived. I realized then that the professor had been inspecting an empty wine bottle in the kitchen. He was probably thinking what I was thinking; she was already sloshed.

"Are your parents still here in St. Louis, Sophia?" the professor asked.

"My father is." Sophie cleared her throat and sat up straight. "My mother died several years ago."

Goldblum shook his head thoughtfully. "I'm sorry to hear that."

She shrugged.

"And what about your parents, Narin?" he asked.

"They're in India right now," I said as I struggled to get the asparagus on my fork.

"That's a place I've always wanted to visit, India," Sophie exclaimed as she clumsily stabbed the asparagus on her plate. "I've heard it's such a beautiful country."

"They are a delightful couple." The professor beamed. "Sophia, I certainly hope you have a chance to meet them someday."

"I hope so, too." Sophie glanced at me. "I hope that's okay with you, Narin." She raised a flirtatious eyebrow, licked her lips, and poured herself another glass of wine. Her cheeks glowed. I took a sip of water.

After dinner and dessert, we cleared the table and loaded the dishwasher with the dirty dishes. I offered to walk Sophie back to her house.

When we reached the street, she grabbed my arm. "Do me a favor. Let's go down to that park." She pointed down the street. "It's not too far, and I could use a walk. And the thought of going home and seeing those disasters I painted today literally makes me want to yak."

"Sure," I said.

"What a clear night." She tilted her head back and raised her arms to the sky. I caught her when she stumbled, nearly toppling over. "Even with all these damn sycamores and all the god-awful light pollution around here, you can still see the stars."

"I like the sycamores," I said. "My street only has oaks and a few sweetgum trees."

"Sycamores are beautiful in the winter when you can see their white bark," she said. "But they shed a lot, and every time the wind blows, it rains bark and those pollen balls. It's the shaggy dog of the tree world." She kicked at the bark scattered on the sidewalk.

"They're not as bad as having those gumballs all over the sidewalk," I said. "I hate those things." I couldn't believe how easy it was to talk to her. I'd never been able to speak to a beautiful girl like that before. Usually, talking to girls makes me feel like I'm taking a test I haven't studied for. But not with her.

"Maples are the best," she said. "They're the right size and shape and turn wonderful colors in the fall."

"I love maples, too!" I exclaimed. "There's one right outside my window at home in Chicago. The leaves turn bright red in the fall."

She grabbed my hand and pulled me toward the park entrance. The park had a grassy field that stretched for a block, dotted with trees. A bright red mat surrounded a swing and a jungle gym. I sat on a bench beside the play area while Sophie hopped onto a wooden beam. She walked with her arms and legs swinging. "I do this sometimes to work on my balance," she said.

I'd done okay so far, but I needed to think of more things to say to her. "After moving here from Australia, where did you live? Did you live in the city?"

"I grew up in a rural area on the Illinois side, where my mum's family is from," she said, hopping off the beam. "It's a small, hick town full of inbred white supremacists. Dad still lives there, but I wanted nothing to do with it after high school." She looked up at the sky. "But living in the country does have its perks. You'd be amazed at how many more stars you see there." She sighed. "My dad considered himself an amateur astronomer. We'd take the telescope and star charts out on the porch to map the sky. He'd grown up in the Southern Hemisphere, so the northern constellations fascinated him. He wanted to know everything about them, where they were, how they moved through the night sky. That was his obsession when I was a kid. Those are some of my earliest memories, sitting with a flashlight and reading coordinates as he adjusted the telescope." She pointed. "Let's play a game. Look over that way and try to guess what constellation that is."

"What am I supposed to be looking for?" I asked.

"Over there, you can see a constellation tonight. For two points, tell me what you see."

"That's Orion's Belt."

"Sweet!" she said, giving me a clumsy high-five. "Now, if you look carefully, you can see the rest of Orion. Oh, I've got something else to show you. Look over there."

"You mean that really bright star?"

"That's Jupiter," she exclaimed.

"That's Jupiter?"

"'The man that lives in the sky.' That's what my dad used to call Jupiter. Oh look, look, a shooting star!" she exclaimed as she jumped up and down. "If you also see one, something terrific will happen to us."

We looked up at the sky, but I didn't see any shooting stars.

"I remember when I was nine or ten, looking up at the sky—" She looked at me. "I'm sorry I'm talking so much about myself, Ravi—I mean, Narin." Even in the dim light, I could see her blush. "Oh, I'm so sorry. I shouldn't drink so much. My mind gets all muddled when I drink red wine."

"Ravi, that's my twin, right?"

"I guess so. Come on, Narin. Let's talk about you. I'd like to know more about you," she said, trying to pull me toward the bench, but I pulled away.

"I'll tell you about myself later. There's something I should have said from the very beginning. I've been feeling guilty about not being completely honest with you." *Here it goes.*

She glanced at me and backed away a step.

"On Tuesday, I start a new job at Harvester Pharmaceuticals."

She considered for a moment. "My dad works there."

"I know. He'll be my supervisor."

She lowered her gaze. The wind blew, and the leaves on the sycamores crackled.

"We had a falling out a while back," she said solemnly. "We don't talk to each other anymore."

"I've only met him once. Maybe—I was thinking—well, maybe I can be helpful to both you and your dad," I stammered. She looked at me skeptically.

We walked back in silence. I wasn't sure if I had ruined my chances with her. I knew I had to be honest. She would find out soon enough.

"You should know one thing about Morey," she said. "He hated Ravi. He was completely opposed to my marrying him. That's why we don't talk anymore."

By then, we had reached my car. Sophie said, "Good night," and went back to her house. There was no hug, no kiss, not even a high-five. I drove away, feeling upset about how the evening had gone. *Shit! What have I done?*

The next day was Monday, but I hadn't started work yet. For some reason, Sanquel also took the day off. I spent the morning in the basement doing squats and practicing my jumps. I finished my last set of jumps: base, heel-touch, and side-to-side jumps, as I'd learned from watching online videos. Then, I lifted weights until noon. The delicious aroma of peppers and onions wafted down from the kitchen. My legs quivered with fatigue as I climbed the stairs.

In the kitchen, Sanquel had pots simmering on every burner. "It looks like you're cooking for an army," I told him.

Sanquel hummed as he seasoned a chicken platter, tossing small pieces to his dog, Chloe, an arthritic, obese black lab. Despite the house rule, Sanquel kept the dog. Chloe sprawled on the kitchen floor, watching Sanquel with her sad and expectant eyes.

I headed to the shower, then went back to help carry plates and utensils to the deck. Sanquel suggested we eat outside since the weather was nice. We brought out the many dishes he had prepared and placed them on the wooden picnic table. Chloe strolled onto a nearby shady patch of grass and collapsed. We piled food onto our plates and started eating.

"My sister, who lives in Shreveport, Louisiana, called me this morning," Sanquel said. I was charmed by the way he pronounced "Shreveport, Louisiana."

He continued, "She's engaged to a quite successful man who manages an oil company. I don't quite know what he does, but he has made a lot of money doing it. I was surprised to hear from her. She called me to tell me that she's pregnant."

"Congratulations." I studied Sanquel's expression. "So, you'll be an uncle soon." His voice had a sharp edge, and I wanted to find out why.

"This man she is engaged to is a —" He furrowed his brow. "He is a Pentecostal. Are those the people who speak in tongues?" He wiped the sweat from his forehead with a nearby towel.

"I don't know," I said. "I don't know much about that."

"The news has upset my poor old mother," Sanquel said. "That is what we discussed. She and my mother had a long, tearful conversation yesterday. My mother wants her to return to Mexico, but she says she cannot—she will not go."

"She's an adult, isn't she?" I was getting angry, but I wasn't sure why.

Sanquel nodded solemnly.

"People should do what they want," I said sharply. "It's her life, after all." I suppose I was thinking about my situation with Sophie, or what I hoped would develop into something soon.

Sanquel poked at his food. "Yes, indeed. But my mother is set in her ways. She is very stubborn and very Catholic."

"But it's not her decision to make," I retorted.

Sanquel sighed. "I agree. Of course, you are right. But I also see it from my mother's point of view. You see, my mother grew up with certain expectations. She always dreamed of doting on her grandchildren, helping them learn their catechisms, and celebrating their First Communion. Now, all these dreams are shattered. It's quite hard on her, and she is no longer in the best of health."

"Yes, but—" I struggled to find the right words. "It's not right for parents to expect so much from their children and grandchildren. It's just not fair to place such a burden on them."

Chloe came over to lick my hand, but I waved her away.

We mainly ate in silence. I muttered a compliment about the food, and Sanquel nodded, clearly lost in his thoughts.

I wished I had something clever to make him laugh or smile, but I had nothing to offer. My mind was blank. He'd made all this fantastic food for me, and all he wanted in return was good company, engaging conversation, and genuine connection. I'd failed to do my part. If he wanted a friend, he'd made a terrible mistake. I've always been a bad friend: always taking without giving back. That's all I could do because, honestly, I had nothing to give. I wasn't fun, attractive, or funny. I was nothing more than a pathetic little idiot. If I were normal, then people would enjoy having me around. But no, I'm just a freak that no one likes to be around. I might as well accept that. Everyone would have been better off if I had gone away to live in a cave somewhere.

When I finish my story, you'll see that the world would have been better off if I'd never been born. But it's too late now. Everything is already fucked up.

Fifteen

On Tuesday, I went to Harvester for my orientation. I wore my only suit again. I bought it five years ago to attend a wedding, but it stayed in my closet until my grandmother's funeral. Now, I was finally getting my money's worth. I took the elevator to the second floor and went to the main conference room. Several other newbies like me were wandering in. We took our seats and waited.

At nine, a woman arrived carrying a stack of thick booklets that she handed out to us. She had a spindly frame and looked painfully thin. Her skin was pale and nearly translucent. Her bright red dress seemed uncomfortably tight, and she walked in a peculiar, sinuous way.

After distributing the booklets, she began her presentation. Her speech was slow and enunciated with exaggerated clarity. "There is a palpable energy at Harvester. Intelligent, hardworking, and passionate individuals work here because they want to make a meaningful impact on the world. You can achieve your professional goals at Harvester while balancing your career and family life. Who says you can't have it all?" It was hard to listen to. She continued, "The leadership at Harvester strives every day, in every way, to be transparent with employees, stakeholders, and the community. The lines of communication at Harvester are always open."

I looked around to see if anyone else thought she was the wrong person to introduce us to life at Harvester. The others just looked bored.

Luckily, I hadn't wasted those twenty minutes listening to her drone on. They had given us a notepad to take notes, and I'd sketched the summer constellations visible in the northern sky from memory. *Sophie's going to be impressed.*

What followed was a film about Harvester's history. Pan flute music played as the company's name appeared on the screen. Sepia-toned footage showed two men in Panama hats and khakis walking alongside naked natives on a jungle path. One was Morey, with his red, cratered nose and thick black hair. The other man looked like Maru—skinny, with large glasses—but had one noticeable difference: an awkward sideways grin. This was Sanjay Chandra, the company's CEO. The natives pointed excitedly at the camera, smiling and laughing.

The scene shifted to an old photograph. Morey and Sanjay Chandra were in the middle, with others leaning into the frame. Everyone was holding up Budweiser cans. Behind them, a homemade sign in block letters read: "Harvester Pharmaceuticals."

The scene then shifted to a medical clinic built by Harvester, where a young, attractive, dark-skinned woman held a baby while a doctor in a lab coat examined her. Inside an elementary school displaying the Harvester logo, smiling brown children sat in neat rows in front of a blackboard. Next, a Brazilian economics professor, with diplomas displayed behind him, explained how Harvester had brought much-needed development to the impoverished people of the Amazon basin. The music swelled again, and the indigenous Amazonians waved and smiled one last time.

The film ended, and another one started. The older Chandra, now gray, explained Harvester's critical mission. I filled out the W-4 as he talked. He stressed the importance of protecting and preserving the South American rainforest from various threats to its survival.

The rest of the day dragged on with varying degrees of torment. There were speakers from equal employment, security, benefits, engineering, manufacturing, marketing, and R&D, each more boring than the last. This was interrupted by a decent box lunch at noon, which included chicken salad on a croissant, a bag of chips, and a giant fudge brownie. The presentations mercifully ended. I signed a form stating that I'd abide by the rules detailed in the employee manual and handed it to the emaciated lady.

We finished around three, and I went up to the ninth floor to see Ian. The middle elevator arrived first, opening directly in front of the grandfather clock. I stepped onto the crimson carpet and felt the strangest sensation that the clock had been waiting for me. I moved closer to it, trying to prove I wasn't afraid of some beat-up, old, haunted grandfather clock. I saw my face reflected in the polished surface of the wood. A vibration from deep inside the clock made the metallic hands quiver.

I reached out and touched the angelic carvings, passing my hand over the worn faces and tightly curled knots of hair. A deep gash cut into the cheek of one of the screaming cherubs. I rubbed it gently, believing my touch could heal this terrible wound. The angel stared back with hollow eyes, its mouth fixed in a perpetual cry. As I gazed at the angel, I could have sworn its mouth gaped even wider, and its tongue protruded just a bit. I was sure it wanted to tell me something, so I leaned in to listen.

"She's beautiful, isn't she?" Ian asked. He had snuck up behind me. "I didn't realize how late it was," he said. "How time flies." Ian moved closer.

I cleared my throat but stayed quiet.

"So, how was your orientation?" Ian asked.

"Fine, I guess," I replied. "I mean, I'm glad it's over."

Ian handed me a key. "Welcome aboard. Morey will want you downstairs, but since the office next to mine is available, I'll give it to you for now so you have a place to relax and unwind if things get too stressful." I took the key.

"I was surprised to learn Morey was so involved in the company's founding," I said.

Ian replied, "He and Sanjay Chandra started Harvester."

"Why isn't he further up the food chain?"

Ian grunted. "Unfortunately, he has a bad habit that has held him back over the years."

"Drinking?" I lowered my voice and looked around. "I guessed that from his appearance."

"You have incredible powers of observation, young Narin," Ian chuckled. "Morey would've been fired long ago if he hadn't been such a key part of building the Neuroscience Division. Luckily, he's not too worried about titles. He owns Neuroscience and is happy with that, though I think Neuroscience could use new leadership." He slapped me on the back. "If you know what I mean." I had no idea what he meant.

I couldn't wait any longer. "Ian," I said, "I need to find out why you wanted me to meet Sophie."

Ian nodded solemnly. "Come with me," he said and stepped into his office. I followed close behind. I took a seat as Ian poured two generous glasses of spiced orange bourbon. We each had a drink. I was starting to suspect that everyone on the ninth floor kept a stash of hard liquor in their office. Morey wasn't the only alcoholic at Harvester.

"I knew Sophie's former fiancé, Ravi Reddy." Ian paused for a moment. "We had become, um, close before his tragic death."

I nodded.

Ian continued, "He was a renowned artist, not just in St. Louis, but recognized internationally." Ian took a drink. "When we were renovating the gallery on the second floor, I negotiated a deal for Harvester to acquire some of his most important works. That's when I met Ravi and became acquainted with Sophie."

"So, you, Ravi, and Sophie were friends?" I asked. "Is that what you're trying to tell me?"

"No, that's not quite accurate," he replied. "Ravi liked to talk about Sophie. She was more than his fiancée. She was his protégé, his project, his legacy. That's how he thought of her. He was working on introducing her to all the right people in the art community. He planned to make her a superstar in the art world. He even asked me to intervene with Morey, to talk to him and see if I could set up a meeting between them. Ravi believed he could sweet-talk Morey into blessing their marriage."

"Did he actually love Sophie?" I asked. "Did he ever tell you that?"

Ian sighed. "I'm not sure how Ravi felt. Love is a tricky thing, you know. But I do know that he was determined to marry her. Perhaps it was all ego on his part, showcasing his young, beautiful, and talented wife to his art friends in New York, London, and Paris. Who knows?"

I tugged hard at my hair, trying my best not to show my anger. "How old was he?"

Ian swallowed. "He was significantly older than Sophie. Mid-fifties, I'd guess."

The tugging was useless. I shook with anger. "I can see why Morey objected!" I yelled. "That's disgusting! He used his fame to manipulate her. He was taking advantage of her." I knew Ravi had died in a car crash. At that moment, I savored a vision of his fiery end.

"Well, she did agree to marry him," Ian countered. "She must have thought she was getting something out of the deal. I'm not sure who was taking advantage of whom."

I started to sputter an incoherent rebuttal, but Ian interrupted me. "You're right. I'm being unkind. Sophie paid her dues." Ian poured another drink for each of us and continued, "After the car crash, she was in critical condition for over a month."

Shit. Here I was, fantasizing about the fiery wreck without realizing that she was in the passenger seat.

"She suffered severe internal trauma, multiple fractures, and a serious head injury," he said. "They kept her in a medically induced coma. They weren't sure she would survive." He let that sink in. "I visited her in the hospital. She was unconscious on a ventilator and didn't know I was there. I did this partly to check on Morey. He was at her bedside day and night. He told me he was praying for her, which surprised me since Morey doesn't seem like a praying man."

I swallowed hard.

"But I could see the truth." Ian appeared lost in thought. "Morey believed that Sophie had gotten exactly what she deserved. He was so full of hate, Narin, it spilled out of him and tainted everything around him. When she recovered enough to go to rehab, Sophie and Morey resumed their bitter feud. They haven't spoken since."

"Okay," I said. "But you still haven't answered my question. Why did you want me to meet her? And, for that matter, how did you know she'd be at Goldblum's house last Friday?"

Ian blinked several times. He held the glass of bourbon lightly in his hand but didn't take a sip. Finally, he said quietly, "Ravi and I were very close. We'd meet up and talk about our work and all the frustrations we were going through. I felt like I'd gotten to know Sophie vicariously. We'd spend hours talking about her. She seemed like a lovely girl who was about to become the art world's darling and share in Ravi's wealth and fame." He paused to collect himself. "But then Ravi died, and she nearly died as well. I kept thinking about her, alone, injured, trying to rebuild her life. Morey would be no help to her. I knew that."

"So, what did you do?" My mouth was so dry I took another sip of bourbon, even though I would have preferred water.

"I did what I could. I'd find ways to secretly help her, pay off her bills, and so on," he said. "To do that, I had to learn more about her and find out who her creditors were. She didn't deserve what happened to her. Over time, she became a project of mine. I wanted to bring her back to the happy world she had before the accident. Ravi would want her to be happy."

"A project?" I exclaimed. "Do you mean an obsession?"

Ian shrugged. "Call it what you will," he said. "I don't apologize for caring about Sophie's well-being. And now that you are here," he said brightly, "I think I can finally make things right for her. I've felt that from the moment I first saw you."

"Me?" I said in disbelief. "What do I have to do with any of this?"

Ian smiled and tipped back his bourbon.

I was shocked by what he told me, but part of me understood what Ian saw. I'm nothing now, yet I was close to becoming one of the most important people in human history. Why wouldn't she want me? Besides, I'm a younger version of her old boyfriend. That must count for something. If Ian believed I was the key to Sophie's happiness, then maybe he was right. Everything depended on my success.

I wandered into the empty lobby. The receptionist wasn't at her desk. An overwhelming sense of loneliness washed over me; it's hard to put into words. I had an uncanny feeling that everyone and everything, except for me, had died. It felt as if I was walking on the black marble floor of an ancient mausoleum filled with the dried-up bones of everyone I'd known.

I left the building, and the door clicked shut behind me. I checked to make sure I hadn't lost my new badge and the key Ian had given me. I found the key in my pocket. My badge was attached to a lanyard I had forgotten I was wearing. The sky was overcast, and the wind buffeted me, rattling the flags overhead and making it seem as if rain was about to fall. I walked to my car and looked back at the building topped with giant block letters: HARVESTER PHARMACEUTICALS. I was now officially part of the team.

I sat in the car, staring straight ahead. "Are you in love with her, Ian? Is that it?" I wished I had asked him that, but would he answer me honestly? Maybe. Maybe not. "What do you want, Ian? What is your game?" I came to my senses, but only for a moment. *You hardly know these people. Just relax and see what happens. Don't think about it for now.* With that reassuring thought, I drove home, rubbing my scalp where I had been pulling my hair. It was tender to the touch. All evening, all night, and well into the next day, my mind kept drifting back to Ian and Sophie. It was a puzzle I couldn't solve.

Sixteen

I arrived at Harvester the next day and went to the conference room on the third floor for the Neuroscience Division meeting. At 10 a.m., Morey called the meeting to order. Each scientist was given seven minutes to give an update on their project. I sat at a back table with Morey.

A poster featuring the company's motto, "Reaping Nature's Bounty for a Better World," hung on the wall behind us. Below the slogan was the Harvester logo, which showed a large green-and-orange combine cutting through a wheat field. A muscular man sat proudly behind the steering wheel of the massive machine, with a copper-colored sky filling the background.

Morey provided a detailed critique of each presentation, concentrating on slide design and layout. He explained, "They call me a slide Nazi, but in my Neuroscience Division, I make sure that people know how to present data. This isn't a worthless academic lab where people who can't speak English present some confused slop that no one can understand." He grew louder and more animated as he spoke. "The slides should speak for themselves."

At the end of the meeting, Morey turned to face me. His watery green eyes looked uncomfortably sharp. I tried not to flinch under his gaze. "JoAnn and Ahmed are part of your team," he said. "Meet with them, and they'll bring you up to speed on the DMTA project. Your office will be next to theirs. Next week, Ahmed will report to your group, but I want you to give us a seven-minute presentation about the work you've done with Cohen. Remember, there's a strict limit of seven minutes, so think about what you want to say within that time frame. You got that?"

"Yes, sir," I squeaked. I wanted an update on Pharma's progress regarding the inhibitor for the Type I receptor modulator site, but I was too afraid to ask. You'll recall that Morey told me Harvester acquired the rights to a drug that binds to the Type I receptor and increases DMTA's affinity for it. He mentioned that the Pharma lab was close to developing a molecule with the opposite effect, inhibiting the binding of DMTA to the Type I receptor. I wanted to know how close they were to having it available for us to work with: days, weeks, or months.

I first met Ahmed Khan and introduced myself. To my surprise, Ahmed responded in Bengali. "You look Bengali," he said, using the formal "you." I struggled to say a few words but quickly switched back to English. Ahmed did the same. He was older than me, probably in his forties. He was short with a protruding belly. He had a gentle, weary voice and a kind face.

We found JoAnn Carpenter in a small conference room. Tall and slender, with thick glasses, she could easily pass for a high school student but was probably in her mid-twenties.

They bombarded me with questions about my work with Cohen. In their eyes, Cohen was a big deal. Finally, they asked the question that was most on their minds. JoAnn asked, "Morey says you have developed a Type II receptor blocker. Is it true?"

They both looked at me expectantly.

"I have," I replied cautiously.

"Holy crap! Does it work?" Ahmed asked.

I nodded. "It completely blocks the binding of DMTA to the receptor. In our mouse model, it prevents the behavioral effects caused by stimulating the Type II receptor. Specifically, the mice don't turn into vicious little killers, regardless of the dose of DMTA we give them."

"Does it actually block the DMTA binding site on the Type II receptor?" JoAnn asked. "Or is it just modulating the affinity of DMTA?"

"It blocks it," I replied.

Ahmed asked, "Would it be difficult for our pharmaceutical team to replicate it?"

I shook my head. "It's derived from a common compound that's been around for a long time. It should be simple." As the words left my mouth, I realized I'd become obsolete once they got their hands on the receptor blocker. They might fire me the next day, and I'd end up with nothing. My heart started to race. *Why didn't I take the money Morey offered? I could end up getting screwed.*

Ahmed interrupted my thoughts. "Is there any insight into why people develop the switch?" he asked. "I mean, the Type II receptors are always present. Why aren't they activated with every exposure to DMTA?"

I shrugged. "Your guess is just as good as mine. It probably has to do with what other neurotransmitters are present at the time of DMTA ingestion. There are plenty of theories, but none have panned out so far."

"So," JoAnn pressed on, "our next step is to get this process to Pharma. They have a good record of promptly providing us with what we need. Morey has already spoken to Chris Vender, and he told Morey that this would be their top priority."

"Tell me about H138," I asked, delaying the inevitable. "Morey said it modulates DMTA binding on the Type I receptor."

They looked at each other. Ahmed said, "It binds to the Type I receptor and increases the affinity of DMTA for the receptor by a factor of ten."

"That's what he told me," I replied. "Morey also mentioned that Pharma was developing a compound that inhibits the binding site and reduces DMTA binding to the Type I receptor." They looked at me blankly when I said this, so I kept going. "He told me that the Pharma lab was close to having it ready. Do you know how close they are?"

They looked at each other again but didn't respond. Their silence made me anxious. "That approach makes sense to me," I said. "You may be able to get the benefits of DMTA without the hallucinations and addiction.

I guess that's what microdosing is all about. But you would then need to block the Type II receptor. It would be a three-punch cocktail." They didn't respond to that either.

"So, Narin, we need a copy of the process for making your Type II receptor blocker. We should try to get it to the Pharma lab today," JoAnn insisted.

I was sweating now. I'd have to hand it over. That was the whole reason they hired me. But I wasn't quite ready. "Out there," I said, pointing to the wall, "there's a rumor that Harvester created Red Sky."

"Oh, God," JoAnn said. "Not you, too, Narin. This is Ahmed's favorite topic."

Ahmed went and shut the door. "Harvester scientists were the first to extract DMTA from savycha. Do you know what happened to the people who did the original work?"

JoAnn shrugged. "Two of them are at Stanford. They developed a side company trying to create an animal model for schizophrenia by messing around with the DMTA receptors."

"I'm familiar with that work," I said, crossing my arms. "I haven't been too impressed with their data."

"What about Carl?" Ahmed whispered. "I interviewed with him when I started at Harvester. What happened to him?"

"I don't know," JoAnn replied. "No one knows. I guess he's not doing science anymore."

"I bet you a million dollars he's the one who developed Red Sky," Ahmed said conspiratorially.

JoAnn looked annoyed. "Don't give Narin the wrong idea. We don't engage in nefarious activities at Harvester. This whole project is about using DMTA safely and therapeutically."

Ahmed looked offended. "I'm not saying he did it on purpose. He might have been kidnapped by a cartel that forced him to develop Red Sky."

JoAnn shook her head, rolled her eyes, and looked back at me. "Okay, Narin. Do you have it?"

I reached into my pocket, pulled out my keys, and unfastened the flash drive. "Everything you want is right here." I held it loosely in my hand. They looked at it with wide eyes, as if it were the Ring of Power. "I suppose since we're working together, there's no harm in sharing the work I've done. It's all in here."

JoAnn reached out, and I closed my fist around the flash drive. "In orientation, they said that we can't use flash drives on Harvester computers," I said. I looked at JoAnn, then at Ahmed, both of whom were staring at me. There was no escape from this. My life was in the hands of Harvester. I handed JoAnn the flash drive.

"Don't worry about what they told you." She winked. "We've got ways around Harvester's security apparatus."

The door swung open, and Morey pushed his way inside. "Are you three having a closed-door meeting here?" he growled.

"We were just telling Narin about the project." JoAnn's voice trembled.

"You've been talking for a long time. I just spoke to the technicians, who told me that you haven't set up the experiments for today yet." Morey's hair was wildly disheveled, and his nose was bright red. "Need I remind you that I want that data by the end of the week? Need I remind you also that this isn't some goddamn academic center where we only have three abstract deadlines a year so you can sit around and bullshit all day? This is the real world, where we, including you three, have performance measures to meet. I also have performance measures. I really hope you don't want me to look bad. Or do you?"

"I'm sorry, Morey. It's my fault," Ahmed replied. "I'll get everything running right away."

JoAnn discreetly slipped the flash drive into her pocket.

Morey continued, "Pharma is working hard to get us what we need." Morey looked my way as if he remembered why he hired me. "Did you hand over the Type II receptor blocker?"

I nodded.

"Good." Morey trudged out of the room.

"He scares me shitless," I said, shuddering.

"Just remember to use Helvetica in 16-point when you give your presentation next week," JoAnn said.

"And no animated slide transitions," Ahmed added. "He hates those."

Ahmed and I each took a diet soda from the fridge.

"So, how did you end up here?" I asked him.

"The usual story."

Ahmed came from a wealthy family in Bangladesh. While pursuing his graduate studies in the US, he fell in love with an American woman, another graduate student working in the same lab. They got married and now have two children.

"Was your family upset?"

"Of course." He shrugged. "They were more upset that she wasn't Muslim than about not being Bangladeshi."

"She didn't convert?"

Ahmed laughed and shook his head. "She's a Catholic, and I'm a Muslim. In other words, we have no real differences in what really counts."

JoAnn and Ahmed showed me around the lab. I scanned my badge, which unlocked the door. The lab resembled others, with black benches housing centrifuges and pipettes. At the end of each bench was a large sink. Several refrigerators and freezers were carefully monitored for their temperatures. There was a sterile, hooded station. The overhead fluorescent lights were harsh, and chemical odors filled the air, but I was used to being in a lab, so it wasn't a problem. They took me to the animal care facility where the mice were kept. This smelled worse, but it still didn't bother me.

My mind kept drifting back to the earlier conversation with Ahmed and JoAnn. They weren't subtle about wanting the Type II receptor blocker, especially JoAnn. They also didn't seem worried that it might be Cohen's intellectual property. After all, I developed it while working in his lab. I decided I wouldn't be concerned either if they weren't worried.

I realized then, as I looked at the caged mice, that it had been over a week since I had a migraine and more than a month since I stopped taking my antidepressants. You'd think that the stress of starting a new job would have triggered a downward spiral of fatigue, disabling headaches, and hopelessness. But it didn't. I wondered if it was because of Sophie. She was the one keeping me afloat now. That's the power of love. Not that I wasn't stressed about saying the wrong thing and ruining my chances with her; that was always at the front of my mind. But, for the first time in a long while, I felt hopeful that things would work out. If I had Sophie and worked toward my goal with DMTA, everything would fall into place.

I decided I would be the first to try the three-hit cocktail once it was ready. I know it's not strictly ethical or safe to use a drug that hasn't been tested on humans, but I was confident that if it worked in mice, it would work on me. Besides, I needed all the help I could get to win her over.

Seventeen

That evening, I called Deepa.

"Are you nuts or what?" I asked.

"I said Rajeev and I were friends," she replied sharply. "I didn't say that we're sleeping together. But you're right. Mom keeps peppering me with all these questions. I probably shouldn't have said anything, at least not until they got home."

"When they get home, they'll want to meet him, the poor guy."

"I'm not worried about that," she said. "They'll like him. He's smart, reads a lot, and enjoys talking about stuff nobody else cares about. In fact, he's a lot like you in that way."

"Let me know when that'll be," I said. "I wouldn't want to miss it."

"Hey, Narin?" She cleared her throat. "Can I ask you a question?"

"Depends on the question."

"Well, you know I've kissed guys before, but not seriously." I could almost see her twirling a strand of hair around her finger, just like she does when she's embarrassed.

"Uh-huh."

"But I've never really kissed a guy, like a long kiss, like making out, if you know what I mean."

I remained silent.

"I mean, I've never really kissed a guy," she said.

"Too much information, Deepa."

"Just hear me out," she shot back. "My question is this: well, is it normal for your lips, tongue, and mouth to feel weird afterward, like for days?"

"I don't know," I said. "I've never kissed a guy before."

"You know what I mean," she replied hotly. "Maybe I'm using muscles I haven't used before." She laughed nervously.

"Don't you have any female friends you can talk to about this?"

"They'd all laugh at me," she replied gloomily.

"You're asking the wrong person. You should know that."

"Yeah," she whispered. "I guess so. I just thought — well, never mind."

"So, what else is happening?" I asked. "How are your med school activities going?"

"I heard pharmacology is going to be a killer," Deepa said. "It's been a while since I've been in the habit of studying. I don't know how I'll get back into it."

"Don't sweat it," I said, laughing. "Most docs don't have a clue about basic pharmacology. That's why they need pharmacists. What else do you take in your first year?"

"I don't know, anatomy, biochemistry, physiology, and maybe some other stuff. I'm getting nervous just talking about it. I'm terrible at memorizing anything. It's like my mind is a sieve or something."

"Anatomy, huh? Do you get to cut up a stiff?"

"Yeah, I guess so."

"Well, I'm glad at least one of us will be a doctor. That should make them happy."

"How are things with you?" she asked. "Do you like your new job?"

"I just started, so I don't really know yet," I said. "Oh, like you, I've also made a new friend."

"Oh, gossip," she said excitedly. "Is she Indian?"

"When have I ever dated an Indian girl?"

"Or any girl, for that matter?" She chuckled.

"Ouch."

"Mom always warned me about the sexual aggression of American women, but that hasn't been my experience at all."

"Yeah, I wonder why," she replied. "So, what's her name?"

"Sophia Whitely," I said. "But she wants me to call her Sophie, which I don't like."

"Why?"

"It just doesn't sound as sophisticated as 'Sophia.' Also, I don't like it when someone's first and last names rhyme. It sounds too sing-songy."

"No wonder you never get any dates," she huffed. "Don't be so picky. So, what does she do?"

"She's an artist."

"Oh," she said. I could picture the wheels turning in her head, and I laughed.

"Is she famous?"

"I don't think so. She sells her stuff locally. I think her work is good, I guess. She's talented and all. But I don't understand how the whole art scene works. Like, how do you even go about becoming famous?"

"So, how'd you meet?"

I heard tapping and could tell Deepa was searching Google for Sophia Whitely. But I already knew the search would be pointless. Hundreds of people online had the names Sophie or Sophia Whitely, but none of them were her.

"It's a long story," I replied.

"What does she look like? Is she cute?"

"I think she's cute. She has shoulder-length dark hair and hazel eyes. She's got a beautiful smile. She's younger than me, in her mid-twenties."

"Send me a picture and make sure she's smiling."

"I need to get one first," I said. "Also, send me a picture of Rajeev."

I had trouble falling asleep that night. When I finally did, I found myself on a high cliff overlooking the red desert. A monkey stood beside the Desert Prince. I recognized the monkey immediately; he was the one sitting on my dad's bookcase. He had jutting eyebrows and a deeply ingrained frown, just like my mother—not the eyebrows, just the frown.

But unlike my mother, he was nearly seven feet tall and covered with reddish-brown fur. Both bowed when they saw me, and I returned the bow, which felt appropriate. My only knowledge of martial arts etiquette came from cheesy martial arts movies. I approached without saying a word.

The Desert Prince spoke, "No two combatants are equally matched in every way," he said. "One combatant will always have an advantage, whether in strength or weaponry. But this advantage is often known by the weaker opponent. For the weaker fighter to win, he must neutralize the stronger fighter's advantage, attack unexpectedly, and keep his own advantage hidden until it's time to use it. This is what you will practice today." I assumed that the weaker opponent meant me. I noticed the knife in my hand. The monkey held a similar knife. Since I had been working out, I felt stronger and more capable of defending myself. But, shit, monkeys are insanely strong.

We faced off, each of us giving a quick bow. I started shifting side to side as the Desert Prince had shown me. The monkey watched me with its intelligent black eyes, assessing me. It lunged for a strike, and I moved to my left, but I was too slow to land an effective counterattack. The Desert Prince made a "tsk" sound. The monkey charged. I lost focus, and my body involuntarily stiffened. It delivered a sweeping kick that knocked me off my feet. Its furry knee pinned down my arm that held the knife. I was helpless. It brought its knife within inches of my throat and smiled. He smelled just like you'd expect a monkey to smell. We sparred for the next hour. I tried different tactics, but they all failed. I felt discouraged. I still had a lot to learn.

The Desert Prince loomed over me. "Your movements are tentative and predictable. Look for opportunities to strike and counterstrike. Seize the opportunity when it arises. A moment of hesitation can prove fatal. Work now on your agility and speed."

I gave them both a deep bow. "Thank you," I said. They both returned the bow.

I'll take this opportunity to talk about these 'nocturnal experiences'; I'm not sure what to call them. You're probably wondering what they were, and honestly, I'm still not entirely sure myself. But they happened—the dreams or whatever they were—so I thought it was worth mentioning them. They didn't feel like regular dreams; they were much more vivid. I felt as if I was standing on that high ridge overlooking the desert, sparring with a monkey. Over time, I started to look forward to these sparring matches and took every bit of advice from the Desert Prince to heart.

The weeks passed as I met various scientists and technicians and became familiar with the lab and its equipment. I attended meetings from different divisions, especially Oncology, which had the most interesting presentations, always had bagels, and served gooey butter cake once a month. Whenever I saw Morey, I asked him when we could start working on the three-hit cocktail we had discussed.

"Pharma's taking its goddamn time," Morey grumbled. "When we have it, we can get started."

I was tempted to ask him why we were conducting so many experiments with H138, the modulator that enhances DMTA binding to the Type I receptor. But Morey didn't seem like the kind of person who would tolerate anyone second-guessing his strategy, so I kept quiet.

Morey was always around the lab. At first, I worried that his irritable style would disrupt the work environment. But once I got used to his prickly exterior, I started to appreciate his approach. Morey could quote verbatim from every article written on DMTA in the last ten years, and he corrected me when I was careless in my characterization of previous research.

Ahmed shared that Morey routinely translated non-English papers about DMTA to ensure that no obscure findings were overlooked. What impressed me even more was that Morey knew every detail of the dozen or so experiments running in our lab. He would get excited about unexpected results, and his enthusiasm would rub off on the team. Moreover, we were in just one lab section. Morey monitored the work of every other section with the same level of vigilance and insight.

One day, JoAnn asked Morey how he got started in neuropharmacology. The four of us sat in the conference room discussing a recent paper.

Morey leaned against the conference table and chuckled wistfully. "I cut my teeth as a young postdoc at Oxford. I worked for a wizened old man who had been involved in developing one of the first SSRIs, Prozac, at Eli Lilly," he said. "This was back in the late 1970s. My work involved taking a slightly different approach. The long and short of it is that we developed the first SNRI." He paused for a moment to let that sink in.

He continued, "We knew we were onto something with that. The university partnered with the Wellcome Trust to run clinical trials. Unfortunately, it was shelved because of liver toxicity issues." Morey nodded solemnly. "But that work paved the way for all the other SNRIs on the market today. We developed all the processes that are still in use."

"Wow," Ahmed said. "Just, wow."

At the Neuroscience Division meeting that week, I gave a seven-minute summary of four years of postdoctoral research. Morey interrupted once, saying, "Look at the amount of space available in the upper left corner of the figure, yet you place the legend on the side. If you moved the legend into that space, into the figure itself, you could make the entire figure larger, making it easier to read. And with the legend there, it's easier for the eye to move between the legend and the graph. This applies to everyone. Look at your slides and ask yourself if you've used your space most effectively."

Ahmed and JoAnn asked a few softball questions, and then I was done.

During the evenings, I lifted weights and practiced my jumps. "I should have seen that kick coming," I muttered as I leaped and landed on the basement floor. By then, I had also added knife work to my workout routine. I was determined to outfight my new sparring partner, Monkey, and looked forward to our next match.

Friday arrived, and I got a text from Ian reminding me about the poker game. When I reached Floyd's house, the players were already settled into their spots, quietly counting their chips. I handed over $200 and took my share. I arranged the chips in front of me and watched Alex discreetly; he leaned back in his chair and slowly twirled a chip between his fingers.

"Let's begin," Ian said, dealing the cards. For the first few rounds, the only sounds were grunts and the clinking of chips as they moved across the table.

"How long will you be in town, Alex?" Carlos asked.

Alex shrugged. "I don't know. My van is in the shop, so I'm staying with my folks until I get it back."

We played several more rounds in silence.

"Didn't Maru use to come to these poker games back in the day?" Alex asked.

Ian chuckled. "I get enough of him at work. Why would I want to see him during my free time?" Ian played another round. "But honestly, Maru's not such a bad guy."

Alex, Chris, and Carlos all laughed, and even Floyd managed a smile.

"I'm serious now," Ian said as he examined his cards. "I know I pile on the contempt as much as anyone."

"There's no denying that Maru is an asshole with no redeeming qualities," Chris said.

Carlos shook his head. "Let us not speak ill of our colleague. I think it's a shame that everyone hates Maru for no good reason except that he has benefited from his position because of his father. But the way you treat him, he must feel very lonely."

"I agree with Carlos," Ian said. "The fact is that Maru is misunderstood."

Chris scoffed. "You can't seriously believe that."

"I do," Ian replied. "When I started working at Harvester, we were dealing with legal issues involving our plant in the Caribbean. Maru and I flew there to sort things out. While we were there, he started opening up to me. You could tell right away that he was incredibly lonely. Like a good lawyer, I let him tell me his side of the story."

"So, what's his excuse?" Chris asked.

"He told me he always struggled to meet his parents' expectations. He didn't want to attend business school, and he had no interest in Harvester. He said that he felt trapped."

"Well," Chris said, "maybe he should leave. He's young enough to go do something else."

"That's exactly what I told him," Ian said. "But he's too afraid to do that, to let his parents down."

"I don't want to speak badly of anyone," Carlos said softly, "but the facts are clear. Maru has no social skills. I've heard that Sanjay Chandra will retire soon. He has been grooming Maru to take over Harvester. How will Maru be able to lead the company when everyone dislikes him?"

"He won't be taking over Harvester," Ian said dismissively. "No one trusts his judgment, not even his father. You may not realize this, but whenever Maru comes up with an idea, Sanjay secretly tells the division heads they can ignore it. Now, tell me, is that any way to run a company?"

"Everyone knows how Sanjay undercuts Maru's authority," Chris said. "Even the janitors." He looked at Alex, who said nothing.

"Maybe you, Ian, should offer to run the company," Alex said. "I remember what you once said about your philosophy on leadership."

Ian furrowed his brow. "What's that?"

"A good leader can persuade his followers to jump off a cliff and convince them that it was their idea all along."

"I said that?" Ian shook his head thoughtfully. "I should write that down."

Eighteen

Sunday dinners with Professor Goldblum became a tradition for Sophie and me. By then, I'd known her for a couple of weeks. She hadn't hinted that she wanted to take our relationship any further, so I didn't push the issue. 'Wait until you're famous,' I kept telling myself. And for the most part, I didn't mind. It was easier to be her friend than her boyfriend. All I had to do was listen to her complain about the weather, the local art galleries, and how ugly popular modern painting had become. She did most of the talking, and I did the listening. The nice thing about listening is that you can't screw it up too badly.

Since wine wasn't served after that first dinner, it was no surprise that Sophie became more reserved. One evening, she announced she had finished the professor's book and called it brilliant. Goldblum looked embarrassed by Sophie's enthusiasm. I didn't mention that I wasn't especially fond of it; it seemed like a jumble of Eastern mysticism and pop psychology. Sophie asked the professor about the regimen of physical and mental exercises he'd recommended in the book.

"I still walk every morning and use the dumbbells daily," he said. "I draw inspiration from the ancient concept of the ideal person. Such a person embodies both sageliness and kingliness. Sageliness requires study and meditation to clear unproductive thoughts. Kingliness demands physical exercise to release excess and unproductive energy. With a calm mind and body, one can conduct oneself with honor, dignity, and nobility. That is what the ancients taught us, and that is what I believe."

I reflected on the professor's words, sensing a connection between what he said and what the Desert Prince had taught me. I'd become weak, both mentally and physically. The Desert Prince exposed my fragile state. But I was taking steps to build myself up. I was finally becoming a man. Now that I was engaged with the real world, I had to prove myself. I knew I could do it; I could make myself strong, intelligent, successful, and respected. I just needed to work hard for both Morey and the Desert Prince. I'd never had a mentor I admired, and at that moment, I had two. I understood how to work hard. Morey was my mentor of wisdom, nurturing my scientific spirit, while the Desert Prince was honing my kingly qualities by sharpening my physical skills. How could I fail with that combination?

Sophie and I went for a walk after dinner, and I pointed out all the constellations I knew. She seemed impressed.

"You've been studying," she said with a smile, leaning against me. This was the first time she had touched me since our initial handshake, and I was thrilled.

"How's everything at Harvester?" she asked.

"Great," I said cheerfully. "Morey is far better than my previous boss. Actually, I think your dad is brilliant. I've never met anyone like him before. He has so much passion for his work."

She remained silent but wrapped her arm around mine.

"How's your work going?" I asked. "I don't even know what an artist does. How do you spend your days? How do you decide what to work on?"

She kicked a fallen branch out of the way. "I've been trying to do all the right things. I've been working with local galleries, trying to put myself out there, which isn't easy for me. My friends tell me I should be on social media and sell all my work on Facebook or eBay. Maybe I could even create some videos and develop a following. I've been working on that. I keep thinking I could develop connections if the right people see my work." She sighed. "But it hasn't worked out all that well. I haven't gotten any traction, and right now, I'm barely making enough to get by." She furrowed her brow. "It's really getting hard. I'm working my ass off to stay afloat. I constantly dread that I could go under at any moment."

I was having trouble following what she was saying while her arm was around mine. But then my thoughts turned negative. If Ravi were still alive, maybe Sophie would be famous and making a fortune. What could I possibly offer her?

"My housemate knows some artists in Chicago," I said. "Maybe he can connect you with them."

She looked at me. "Yeah, maybe. I think the main thing I need to do is get over my anxiety. My painting technique is fine, but I'm constantly worried I'm not quite getting the picture in my head onto the canvas." She had a look of desperation on her face. "Do you know what I mean?" She kicked again at the branches scattered on the sidewalk and continued, "You know what Voltaire said, 'The perfect is the enemy of the good.' They say that perfectionism goes hand in hand with fear. I guess that's why I spend so much time on everything. I don't want people to think I'm just a hack. I want to create things I'll be proud of. I know I shouldn't compare myself to others, but I keep feeling like I'm falling further and further behind where I want to be."

I searched for the right words to comfort her, but nothing came to mind. So I said, "I didn't know Voltaire said that." Then I just felt stupid. We walked along, each lost in our thoughts. She knelt to pick up some leaves for a project. Some were freshly fallen, while others were withered and cracked. She looked captivated by the texture of these fallen leaves.

I reflected on how different her life was from mine. She experienced all the colors and shapes of the world around us, while my existence was rooted in chemical formulas and PowerPoint slides. Vitality flowed through her, a vitality I lacked. I could learn a lot from her, like how to live, appreciate, and love. I had just met her and already found myself smiling unexpectedly, like someone recalling a funny story or a juicy secret.

My heart was pounding. I needed to act so she would understand how I felt about her. Under a solitary maple with silver bark and fresh green leaves, I drew her close and leaned in to kiss her. She gently pushed me away, leaving us both visibly embarrassed.

"I'm sorry," we both said simultaneously.

"I was just—" I tried to think. "I'm so sorry."

What was I thinking? I fucked that up. I really fucked that up.

"It's okay," she said with an awkward laugh. "I was just surprised, that's all."

We walked back to her place in silence. She clutched the leaves in her hand.

We approached her house and stood under the yellow porch light. I asked for permission and took a picture of her. "For my sister," I explained. "My sister wants to know if I have made any friends." I was worried it was too dark, but the porch light lit up her face. I hoped she would smile for the picture, but instead, she looked troubled. I took the photo, turned to leave, but she called me back.

"Not here," she whispered. "Let's get away from the light."

We slipped into the shadows. Our lips touched gently for just a moment, then it was over. I should have felt ecstatic, on cloud nine, but I didn't.

She furrowed her brow and looked at my face. "What's wrong, Narin?" When I didn't respond, she kept going, "Don't worry, everything is going to be fine." We had crossed a point of no return, and we both knew it.

"I need to know something, but I don't know how to ask." My eyes drifted into the darkness. I licked my lips. "Maybe we should discuss it over wine next week."

"No," she said sternly. "Let's talk now. What would you like to know?"

"I met someone named Alex Lifer at a poker game." I shifted from one leg to the other. I don't know why I was bringing this up, but now, after making so much progress, I was heading straight into dangerous territory. I needed reassurance that there wasn't something in her past that could come back to bite me.

Her face flushed, and her eyes narrowed. She pressed her fingers to the bridge of her nose. "That's not a question, Narin. What do you want to know?"

"I was at a poker game. Someone mentioned a sordid affair between you and him—and then everyone at the table laughed." I started tugging at my hair. *What the hell am I doing? Just let it go.*

She threw up her hands, and an exasperated sigh escaped her lips. "Let's have a drink. Come on inside."

She unlocked the front door, and I followed her inside.

From the outside, her house appeared to be a simple, white ranch-style home. We stepped into a spacious living room with wooden floors and a tall, arched ceiling. Moonlight poured through the skylight above, casting uneven patches of light across the whitewashed walls. She turned on the lights. Art supplies were spread out around us, and the scent of paint and solvents filled the air. Sophie moved unsteadily from one painting to another, covering each with a paint-stained sheet. "I don't like people seeing them until they're finished," she said.

I was worried that all the chemical smells would give me a headache, but strangely, they didn't.

She asked if I was allergic to cats, and I told her I wasn't. She explained that she had two: Pablo and Toro.

She grabbed a bottle of Chardonnay from a rack and poured two glasses. We settled on the floor next to the couch, drinks in hand. I'm not great at sitting on the floor, so I leaned against the sofa for support.

"I didn't realize Alex was back in town," she said, gulping down her glass of wine. The bottle sat between us. "We'd been seeing each other on and off since art school. He's a pothead and a bum. I don't know what he's up to now, but I doubt he's changed much. I'm sure he's a Red Skyer like all the rest of his type. I haven't kept in contact with him since we broke up. We didn't have much in common back then, except that we both enjoyed outdoor activities like mountain biking, climbing, and spelunking. I don't think he cared much about art; school was just an excuse to avoid real work."

She poured herself another drink. I hadn't touched mine. She continued, "I guess toward the end, he was more serious about me than I was about him because when he found out about Ravi—" She rolled her eyes.

"He totally flipped." She took another drink. "I tried to let him down gently, tell him that I needed to move on. All of a sudden, he started saying that he couldn't live without me. He wanted me to move to California with him and live in that smelly van." She shuddered.

I observed her every move with an intensity I'd never felt before. "What was the sordid affair?"

She swayed as she spoke. "He made the mistake of going to Morey's house because he thought I was staying there. Maybe I told him that. I don't remember." She finished her glass and poured another. "Alex came up to the house screaming and making demands. Luckily, I wasn't there." She made a gurgling, high-pitched laugh. "Well, Morey wasn't having any of that. He came out with a shotgun, and according to the police report, they exchanged a slew of colorful insults. When Morey couldn't get rid of him, he blew a hole in the windshield of Alex's van."

"Morey did that?" I stared at her, my mouth dropping open in shock.

"He got arrested, too." Sophie finished her drink. "'Improper discharge of a firearm,' though I'd argue it was perfectly proper." She giggled. I watched her languid movements as she arched her neck and back. She continued, "After that, Alex disappeared from my life. Morey got a year's probation. And—" She raised an eyebrow. "I got Ravi. Aren't you going to drink?" She nodded toward the glass I was holding.

"I would have loved to see Alex's face when Morey blew out his windshield," she said, shaking with laughter. "He loves that van more than anything else in life. Oh, I wish I could've seen his face!" she howled with delight.

I took a sip of my drink. "How did you meet Ravi?"

She rested her head on the couch. "Ravi was my painting teacher, ironically enough, the very one who said that I didn't have the talent to become a professional artist."

"You were dating your teacher?"

"It's not as bad as it sounds," she said, waving a wayward arm. "I'm not the kind to screw my teachers for a good grade. There were plenty of people in my class who were doing that, but that wasn't me. The fact is, he was my bitter enemy when I was his student. But then I met him socially," she added, and pointed at me, "at an event sponsored by Harvester, where they bought some of his paintings for their gallery. He turned out to be cool when he wasn't teaching. We started seeing each other socially, as friends, and one thing led to another."

"But Morey didn't like that." I studied her face and gestures carefully. I sensed there was something I needed to uncover, something she wasn't revealing to me.

"Morey went ballistic!" She mimed an explosion with her hands. "He hated that Ravi was so much older. Then there was the race thing. Morey kept talking about how he understood immigrants and their ways. He grew up poor in Sydney and was surrounded by 'those foreigners.' He said they're all sweet and kind when they're romancing you, but it's a different ballgame once you belong to them. They can't help but imprison their women. That's just their culture. It's the same bullshit I'd heard all my life. Finally, I couldn't take it. I cut Morey off. I didn't return his calls. I wasn't going to let him tell me what to do. After that, Morey dove into his worst behaviors. He drank and gambled. He used my supposed treachery as an excuse to distance himself from his friends. He said I had ruined myself and humiliated him. He claimed he didn't want anyone to see what my betrayal had done to him. Ravi—that guy has some balls—even tried to talk to my crazy father. He was convinced that if Morey got to know him, everything would be fine. But Morey wouldn't even respond. He just walled himself off."

We sat quietly, and I took another sip of my drink.

Two years ago, just two weeks before we were supposed to get married, we got into a car accident. Ravi tried to stop—" Her eyes widened, and her voice broke. "At first, Morey tried to support me as I recovered. But he was too far gone. He was consumed by bitterness and anger, driven by hatred. He believed that Ravi and I got exactly what we deserved." Her voice grew thick, and tears gathered in the corners of her eyes. I moved closer and wrapped my arm around her as she sobbed for several minutes.

"I told him to leave me alone," she said between sobs. "And he's left me alone. I've been on my own ever since." She took a deep breath, trying to regain her composure. "I used to have so many friends, but they've all moved on. I used to love going out and doing things, but now all I do is work. It's all I can do to keep myself afloat. Every month, I worry I won't be able to pay the rent. I don't have time for friends anymore." Tears rolled down her face. I held her tighter. The scent of her hair and her warmth sent me into ecstasy. But the question running through my mind was: Why did she agree to marry Ravi? Was success so important to her that she would give herself to him? And what does that say about her priorities, character, and our future together?

A big black cat crawled out from under the sofa onto Sophie's lap. She smiled. "Narin, I'd like you to meet Toro."

Nineteen

After about six weeks at Harvester, I had established a routine. It was now deep into June, and the days were getting longer and warmer. One day, I decided to shake things up. I started my morning on the second floor. No one was there. It was eerily silent. I went to the entrance of the art gallery. There wasn't a door. The stark white wall curved into a dimly lit room. At the entrance, I read the plaque: 'Ravi Reddy finds inspiration in the dreamlike quality of Indian cinema. He uses vivid colors and broad strokes to bring his figures to life on the canvas.'

Someone needed to change the plaque to the past tense.

I examined each of his paintings. Some were on a single canvas, while others spanned multiple panels. Swans glided over a translucent lake beneath the moonlight. Almond-eyed girls posed seductively on the shoreline. Two figures intertwined with flowing, elongated arms and legs. The colors were vivid and dramatic.

My mind quickly started to drift. I had been training every day and sparring with Monkey every night. I was becoming stronger, faster, and more confident. I had gotten used to how he moved. I knew when he was about to strike. Still, I kept losing. I was getting closer, but I wasn't there yet. I still didn't know what I was preparing for, but I completely trusted the Desert Prince.

I shifted my focus back to Ravi's paintings. He was an incredibly talented artist, and I knew I could never match his skill. But what did his art truly mean, if anything at all? I tried to understand the person behind the artwork. The more I studied his paintings, the more I felt I was missing something important. I couldn't see it, but I knew it was there.

Ravi had two clear advantages over me. Ravi was both incredibly talented and completely dead. The dead can't do anything wrong. Unfortunately, I was so prone to messing up—saying the wrong thing, misreading feelings, or just being a jerk—that I risked losing Sophie every time I talked to her.

But despite the risk, I visited Sophie's house several times a week after work. She'd paint, and I'd tell her about Harvester. I'd describe whatever tirade or tantrum her dad had that day. She laughed and said, "That's Morey, all right." I was settling into a routine: having morning coffee in Ian's office, working all day in the lab, visiting Sophie's house some evenings, having dinner with Sanquel on other nights, playing poker on Fridays, and having dinner with the professor and Sophie on Sundays.

After leaving the gallery, I went upstairs and started my work. At first, JoAnn and Ahmed seemed happy to have me on the team, and I was making steady contributions. However, by the beginning of June, I noticed that JoAnn and Ahmed appeared to avoid me. They were polite, but they always spoke to each other instead of to me. They rarely asked for my opinions on the results of the experiments or even how I was doing. Admittedly, they were busy with their part of the project. Having worked together for a long time, they felt more comfortable with each other than with me.

I also noticed strange looks from them as they whispered to each other in the corner of the lab. Maybe they thought I was no longer needed. They didn't say it out loud, but they didn't have to. I wasn't even sure if they trusted me or believed I was capable of doing the work I was assigned.

I tried to tell myself I was imagining things. Then, one day, I saw Ahmed looking at my notebook. I kept my notes in a notebook and transferred them to the project site on Harvester's system at the end of each day. Company policy strongly discouraged this, but I was used to keeping a handwritten lab book. When Ahmed noticed me staring at him, he closed the book, even though it had been open on the table. He laughed nervously. "I'm just making sure that our work isn't overlapping," he said, quickly walking away.

One day, I ran into Carlos and Chris in the cafeteria. I'd seen them every Friday at the poker games, but I'd never seen them at Harvester. They sat at the end of a bench, eating sandwiches. I brought my tray over and sat next to them, thinking it was a good chance to gather some information.

Carlos greeted me warmly and asked how I liked Harvester. He wore a large gold cross on a chain around his neck. Chris nodded in acknowledgment as I sat down, but he didn't say anything. After exchanging a few pleasantries with Carlos, I asked the most pressing question on my mind.

"So, Morey tells me that you're very close to having a modulator that inhibits the H138 binding site on the Type I receptor," I said. I waited for a response, but when they said nothing, I continued, "I'm eager to get started on that. If we inhibit DMTA binding to the Type I receptor, we'll be able to develop a drug that prevents hallucinations and addiction, making DMTA a viable therapeutic option. Morey thinks it'll be a real blockbuster."

They exchanged glances but stayed silent.

"So, when do you think it will be ready?" I asked as casually as I could.

Chris spoke. "There are only two things we're working on for your group," he said flatly. "We are working on replicating the Type II receptor blocker you brought from Cohen's lab. The second thing we're doing is trying to speed up the production of H138. We are refining the manufacturing process to make it easier to produce. Both projects are progressing according to a schedule acceptable to Morey." He took a bite of his sandwich.

I was stunned by this and looked at Carlos for confirmation. He looked away. "But why are you working on mass-producing H138? It increases the binding of DMTA to the Type I receptor, amplifying the hallucinations and euphoria. That's the exact opposite of what we want."

"We're following orders," Chris replied dismissively.

"But what about the inhibitor?" I asked desperately. "Is that even in the pipeline? Morey told me that you were very close to having it ready."

Chris looked up at me for the first time. His eyes seemed to ask, 'Are you really that stupid?'

"If Morey told you there was an inhibitor of DMTA binding at the Type I receptor, then he was lying," he said. "There is no such thing, and with our current workload, it won't be in development for at least another six months." He glared at me. "Don't you realize that Neuroscience isn't the only division at Harvester?"

Carlos interrupted. "Narin, listen." He paused briefly to gather his thoughts before continuing. "Morey often hides the true nature of the projects in your division. Corporate espionage is a serious issue, and he has trust problems."

"But why is he asking you to mass-produce H138?" I asked, feeling unsteady and sick. *What on earth were they telling me?*

Chris finished his sandwich and crumpled the wrapper. "That, Narin, is a question you should ask your boss."

Carlos nodded, and they walked away, leaving me completely confused. Everything I believed about our work at Harvester turned out to be a lie.

I couldn't bring myself to return to the lab. Instead, I went to the gallery to look at Ravi's paintings again. I still felt as if the paintings were trying to tell me something, but I was too dense to understand.

Twenty

The rest of the week blurred together. I kept trying to understand what Chris and Carlos had told me. If JoAnn and Ahmed knew anything, they weren't sharing it. Honestly, I was too afraid of Morey to ask him. Twice that week, JoAnn and Ahmed hadn't shared details about an experiment in the lab, and Morey had come down hard on me for 'not knowing what the hell was going on.' Later, Ahmed apologized and said it wouldn't happen again.

The best part of my day was visiting Sophie's house. I watched her sharp eyes scan the canvas, soaking in the scene and planning her next move. After she finished, she covered all the paintings she had worked on. I wasn't sure why. What did she not want me to see? But it didn't bother me.

We might whip up a quick meal in her kitchen or go out for a snack. We took long evening walks, complaining about the heat and humidity. We visited museums, the zoo, or hiked in the woods every weekend. Sometimes, we kissed, but we didn't go beyond that. I was comfortable keeping it that way, and she didn't seem eager to take things any further.

On payday, I logged into the HR website to check my electronic pay stub, which showed a deposit of $4,268 in my bank account. I reviewed each line, including federal, state, Medicare, and Social Security taxes, as well as life insurance, health insurance (with good, gold-level coverage), a 401(k) plan, and other benefits. I felt grateful for a steady paycheck for doing something I'm good at. While things weren't perfect, they could have been worse. I no longer had to worry about grant proposals or manuscripts being accepted, nor did I have to stress about collaborators dragging their feet or backing out at the last minute.

I just needed to keep my eye on the ball and develop the cocktail to make DMTA a drug everyone would want. This would be my contribution to making the world a better place. I committed myself to staying on Morey's good side to accomplish this. He had an explosive temper, but after his outbursts, he returned to his usual self as if nothing had happened. He'd bring techs to tears and then ask them how their grandmothers were doing.

The hours were long, filled with planning experiments and analyzing data. But we had a small army of techs handling the challenging work of running the experiments. Sometimes, I was breathless because everything seemed to move so fast. Morey always stayed nearby. I couldn't understand how Morey managed to spend so much time with us while keeping an eye on the other neuroscience groups. Maybe he had created a team of clones, another Harvester secret.

Earlier that week, I shared the preliminary results of a series of experiments with H138 with Morey. Morey was thrilled, and I felt like I was on cloud nine for the rest of the day. I just wished I knew what Morey was planning. I tried to tell myself that Chris and Carlos had lied to me.

The other thing that puzzled me was Ian. I needed to talk to him again. After finishing my work one evening, I went to the ninth floor. The door to Ian's office was slightly open. Ian was holding tweezers and gently dropping a wriggling creature into the bird's mouth. When he saw me, he gestured for me to come in.

"I've made an important discovery, Narin," he exclaimed. "She loves mealworms." Ian waited for the bird to open her mouth again. "Which is a good thing, too. She's gotten tired of the crickets." He picked up another mealworm. "I'm thinking about visiting your lab and grabbing a mouse. They like killing and eating small rodents."

"Just don't get one with DMTA in it, or you'll have one hell of a problem."

Ian chuckled. "How can you tell if a rodent has been exposed to DMTA?"

"It has a small but noticeable twitch," I replied. I watched Ian feed the bird. "You look tired," I said.

"I haven't been sleeping well," Ian replied, picking up another meal-worm. "Ever since I started working here, I've had trouble sleeping." The bird snapped up the worm and squawked happily. Ian's face lit up. "So, how's Sophie?" Before I could answer, he asked, "Does Morey know about the two of you?"

I ignored the question. "Sophie told me she started dating Ravi after meeting him at the gallery's dedication downstairs," I said.

Ian hesitated briefly before picking up the next mealworm. "She attended the gallery opening," he replied. "I remember it well. She was wearing a short black dress, and her hair was styled in curls. She looked quite charming. When I approached her to introduce myself, she complained that it was too cold, as if I were responsible for the temperature in the hall. I imagine that a girl like that is always cold. Beautiful girls usually are."

"Are what?"

"Cold," he replied as he placed another mealworm in the bird's mouth.

"I don't think she's cold."

"I'm glad to hear that," Ian said, setting down the tweezers. He closed the lid on the mealworms and turned to the bird. "You're full, aren't you?"

"There's something I'd like to discuss with you," I told him.

"Good. I wanted to talk to you as well," Ian said as he approached his desk. "Look at Sophie's painting. As you can see, I've added some proper lighting." He pointed to the lights overhead.

"The mangos?"

Ian stood in front of the picture, and his shadow fell on it. "I've been studying it with scholarly intensity over the last few days, and the only conclusion I can draw is that Sophie is an extraordinary woman. Despite her coldness, or maybe because of it, she needs more than just an ordinary man."

"Is that right?" I had to laugh. "You've come to that conclusion just from studying her painting?"

Ian nodded thoughtfully. "I can see that you're skeptical of my scholarship."

He came over and sat down next to me.

"Don't become complacent," he said with unexpected seriousness. "I told you to try to become her friend, and things would develop naturally. But now it's time to take the next step."

"I have no idea what you're talking about," I said, laughing.

"But it's essential that you understand." He lowered his voice. "Ravi always complained that Sophie was restless," he said. "She would get bored and create her own drama. She found Alex and his nomadic lifestyle enticing until she didn't. She reveled in the glamor of Ravi's fame and artistic success and pursued him so she could be part of it. I'm trying to say that you could lose her if you're not satisfying whatever fantasy she cooks up in her mind."

I felt a chill run through me. Was it that obvious? How did Ian know my biggest fear was that Sophie would think I was a bore and lose interest? Here I was, on the verge of changing the world. I was so close! Once I gained fame, Sophie would be mine. I couldn't let Morey push me around anymore. This was more important than me, Morey, or Harvester. We needed to get back on track with the project I had in mind, which Morey had told me about on that first day.

Ian interrupted my thoughts. "She can't get Ravi back," Ian said. "But you are a perfect substitute if you can prove yourself worthy."

"A substitute?" That rubbed me the wrong way.

"I don't mean it like that," Ian said. "I mean—the poor girl has been through a lot. But I believe you can help her overcome her sorrow. And the best way to position yourself to do that is to stick with me." He stared at me with bloodshot eyes. "Just do what I say, and I'll make sure you quickly climb the corporate ladder and receive all the credit for whatever DMTA concoction you create. You said you knew how to use it as a cognitive enhancer and an antidepressant. I'll ensure you have all the tools you need to do that." His voice became hoarse. "I'll be the best damn friend you ever had. I'll stick by your side no matter what happens."

I still didn't like him calling me a substitute, but I didn't press the issue. "Believe it or not, I'm not here to talk about Sophie," I said coldly. "I was hoping that you could tell me the truth."

He stood up and took two shot glasses and half a bottle of vodka from behind his desk.

"The truth is like an onion, Narin," Ian said and sat down beside me. "If you peel away the layers, it will only make you cry. Think carefully. You're in the middle of a critical project, and very few people here understand what it's all about. Are you sure you want the truth?"

"Morey lied to me," I said, pointing out the obvious.

He nodded. "Shit," he said. "Chris and Carlos told you, didn't they?" Before I could reply, he added, "I should've thought of that before inviting you to the poker game. That's my fault."

"Do you know about H138?" I asked.

Ian gave a quick nod.

"It increases the binding of DMTA to the Type I receptor," I explained. "It amplifies the effect of DMTA. I don't want to get into all the details."

He replied, "I know the details better than any lawyer should."

"When I first met Morey, he told me that Pharma was close to developing a way to inhibit the H138 binding site, which would lead to much less DMTA binding. If that were true, it would be a major breakthrough. Such a drug could eliminate concerns about hallucinations and addiction while preserving all the positive effects of DMTA. That would be a significant achievement. Also, if we had a Type II receptor blocker, people could enjoy all the benefits of DMTA without worrying about falling into the switch."

"I know all that," Ian said calmly.

I yelled, "Then why is Pharma working on a way to mass-produce H138?"

Ian took a deep breath. "Because Harvester does more than manufacture pharmaceuticals to treat diseases."

I didn't understand what Ian was saying. I sat squirming in my chair, wondering what he meant.

"So, you want me to peel the onion?" Ian asked. "Well, in that case, we'll need a knife."

Ian filled the two shot glasses with vodka. "Drink up." I looked at my shot glass suspiciously, but after Ian downed his glass, I did the same in one swallow. Ian laughed. "That's how to drink a shot of vodka," he said. "When I first joined Harvester, Chris, Carlos, and I would often go out for drinks on Fridays. Floyd would sometimes show up. Maru would tag along. While we were taking shots, Maru would sip from his glass as if he were having his afternoon tea." Ian laughed, grabbed the bottle, and poured another round. "Have you ever known anyone who sips from a shot glass?" Ian shook his head. "That's why we started having our poker games on Fridays so that we could ditch Maru."

Ian finished his second shot and took a deep breath. "I expected you'd come to me eventually, so I've been thinking carefully about what I should say." He sighed. "Morey brought you on board because he thought you had secretly developed a Type II receptor blocker in your old mentor's lab," Ian said.

"He was right," I replied. "But there's no reason to keep me in the dark. I'm not an idiot. I know what we've been working on for the past few weeks. It all revolves around using H138 for—something."

"Harvester acquired the rights to H138 for a specific reason." Ian squinted as if seeing me for the first time. "Can I trust you, Narin? Can I really trust you?"

"Sure," I said. I was being flippant, but I didn't care.

"No." Ian slammed his shot glass onto the table. "Don't say it like that. Say it like you mean it. I need to be able to trust you."

"Of course, you can," I said. "I give you my word of honor. After all you've done for me, I'd do anything for you." And I meant it. I couldn't help but feel affection for Ian. He got me this job, loaned me his car, and set me up with Sophie. At that point, I could honestly say that Ian was the best friend I'd ever had.

"Good," Ian said. "I'm glad. I feel like I've been deceiving you since the moment we met. I want to clear the air. I don't want secrets between us."

Ian poured another drink before getting up. I still had my second shot glass full. He carried his glass and the bottle of vodka with him.

"Come with me," he said, then walked to his desk, and I followed.

"I have something to show you," he said.

I stood behind him as he typed on his computer. A grainy video appeared.

"An interesting thing happens when you give H138 and DMTA together, not to mice, but to humans," Ian said.

"What am I looking at?"

"A recent interrogation video."

The video showed a brown, bearded man strapped to a chair. In the top right corner of the screen, a banner read: "OPERATION JUGGERNAUT." A few people moved in and out of the frame, ignoring the man's pleas for release. Subtitles appeared at the bottom of the screen: 'I'm innocent. Please. Please.'

One of the men in the video said in English, "Five milligrams," and injected a syringe into an IV in the man's arm. The bearded man struggled to break free from the straps.

After twenty seconds, he ceased struggling. His eyes started darting around, and he seemed to be hyperventilating.

"Who is your commander?" a man asked in English, and an interpreter repeated the question in Arabic.

The man remained silent.

The English-speaking man said, "Give him another five milligrams."

The process was repeated. After administering a third dose of medication, he answered the question.

"There," Ian said, bursting with excitement. "Did you hear that? That was one of Iraq's most notorious terrorists, responsible for the deaths of dozens of US soldiers."

"Where is your camp located?" an off-screen voice asked. This was translated again into Arabic.

The bearded man responded emotionlessly, as if discussing a trivial matter.

"They got him," Ian said. "That night, they bombed the terrorist camp and killed fifty enemy soldiers while they were sleeping. And not a single American life was lost in the process."

The bearded man suddenly lunged forward in his chair as if trying to bow before God. He let out horrifying screams that I felt in my chest. He was losing control. "Holy fucking shit," I said. "What's happening?" My heart was racing. I had to rest my hand on the desk to stay steady.

"One of the unfortunate side effects of this combination is that everyone always goes into the switch at eight minutes after the injection," Ian replied blandly. The man kept thrashing, pulling at his restraints.

"Why aren't they giving him DMTA?" I yelled. "That should abort the switch?"

Ian shook his head. "For some reason, that doesn't work in this situation. The switch is permanent. They have no choice but to euthanize the subject." He pressed a key, and the video turned off.

Ian stood up and patted me on the shoulder. "This could turn the tide of the war. You've seen the crude methods the military has used to gather intelligence, beating people to a pulp and all the atrocities that happened at Abu Ghraib. They have no idea what they're doing, and they know it. They're desperate to get actionable intelligence. And what we have here in this little magic cocktail is the solution to their problem. It eliminates all the suspect's inhibitions and makes him compliant." He reached into his desk drawer and pulled out some papers.

"This is a memo we received from one of their scientists based on their initial tests," Ian said. "Morey says he's one of the top psychiatric researchers in the country, and he sounds pretty damn excited about the whole thing." Ian read the memo to me. The words were etched in my memory forever.

The DMTA/H138 drug combination seems to affect the same neural pathways involved in hypnosis and dreaming. The subject's experience includes vivid imagery, reduced inhibition, and diminished rational control. The sense of self and surroundings becomes fluid, and importantly, with DMTA/H138 present, the subject becomes susceptible to the interrogator's suggestions. He is placed in a waking dream. The interrogator can shape the experience, ask questions, and direct the subject to reveal whatever he desperately wants to hide. The subject cannot resist. It's as if an unimaginable, God-like force—a Juggernaut—has confronted him. The interrogator holds that power. The only option for the subject is to surrender completely. From a scientific perspective, this presents a valuable opportunity. For the first time, we are gaining a clear understanding of the connection between dreams, mystical experiences, and psychosis.

"But eight minutes of clarity isn't enough," Ian said. "That's why we needed a Type II receptor blocker. We need to stop the switch from happening too quickly."

I nodded and opened my mouth to speak, but I couldn't come up with anything to say.

Ian grabbed my arm. "You've done something Pharma has spent a year trying to do," Ian said. "Chris is pissed that you showed them up like that. Completion of this project wouldn't have been possible without your help."

I still couldn't speak.

He moved closer. "You should be very proud of yourself, Narin." He adopted a more serious tone. "This entire project, Operation Juggernaut, is classified, so we can't disclose it to anyone. But once we have an inhibitor for the H138 binding site, we'll have a product we can proudly share with the world. And who wouldn't want to use it if it makes you smarter and happier?"

I was struggling to breathe. "But you can't— The Cranston-Cohen model is just a hypothesis. It hasn't been substantiated. Morey even called it bullshit the first day I met him."

He looked at me with those sky-blue eyes. "But you think it's true."

"But—how can you be sure?" I pleaded.

"We'll find out soon," Ian said. "The tests on other enemy combatants in Iraq and Guantanamo Bay will tell us if the theory is right or not."

Before I could say anything, he added, "Two hundred and fifty million dollars is on the line. But that's only the beginning. If this works, it could bring in billions for Harvester. After all the bad news we've had, with the lawsuits and everything else, this will be the thing that saves this company."

I stood there in shock. This was totally beyond what I had imagined.

"I spoke to Chris today," Ian said. "They'll have your Type II receptor blocker ready by next week."

He poured himself another drink and looked at me sympathetically. "This is big, Narin, really big. We're trying to get the truth out of evil people," he said and pointed to the screen, "from criminals and terrorists, and we're trying to do it without resorting to barbarism and torture."

I turned away from the screen and slowly made my way toward the door. As I moved, it felt like my body was falling apart, shedding pieces. Only my shadow remained behind.

"Narin, come back," Ian yelled. "We're not finished."

I pressed the elevator button, but the light stayed off. Behind me, the wide-eyed, open-mouthed angels on the grandfather clock stared at me. I wasn't going anywhere. I returned to Ian's office and sat down.

"You gave me your word that I could trust you, and you also signed a confidentiality agreement," Ian said softly. "I know you're a man of your word."

I nodded.

"That agreement covers everything you saw tonight. It would be a terrible idea to break that agreement," Ian said sternly. "Do you understand that?"

"Yeah, I get it," I said. "You don't have to threaten me."

Ian laughed. "I'm not threatening you. I don't threaten my friends. I'm giving you advice because I like you. We're a team, you and me. We're partners." Ian smiled, but the smile faded.

"Thanks," I muttered.

"Maybe I made a mistake in showing you this," Ian said. "I did it because you asked. I wouldn't have done it if I didn't believe you would do the right thing."

I nodded. "Can I go now?"

"Harvester needs you," Ian said. "And I need you."

I looked at him. I loved Ian for what he had done for me, but I also hated him for trapping me as he had.

I didn't leave immediately. I went to the second-floor gallery and examined Ravi's paintings. What was I missing? There was something there.

And then I saw it: a red moon, a red dish on the table, the red bindis on the women's foreheads, the bright red eyes of a raging lion, and a warrior holding a round red shield.

I approached the last painting, which showed a dark bedroom. Crimson light streamed through the window, illuminating the scene. The bed was unmade, and several chairs were overturned. From a toppled wicker basket, beer cans and cigarette butts spilled out. A small revolver sat on the bedside table. I knew it was there, so I examined every inch of the painting. And there they were, the unmistakable red circles. They glowed in the crimson light. Two small red pills rested in an ashtray next to the revolver.

I couldn't help but smile—a dark, sinister smile.

"Now I know a dirty little secret about you, Ravi." An icy thrill shot through me. "You're not perfect after all. You're a fucking Red Skyer."

Part III

Twenty-one

Officer Coffin entered the interrogation room and watched me scribble away. He asked if I was almost finished. I told him, "Not quite." He probably thought I was crazy, writing page after page like a man possessed. I'm sure he dreaded having to read everything I'd written. He said a psychologist would come tomorrow to evaluate my "mental state."

He and a guard took me to my cell. I brought my stack of paper and a pen. The guard didn't seem interested in talking, so I didn't ask for his name. I usually don't ask people their names, but Red Sky made me feel unusually social.

Officer Coffin told me that, besides the psychologist, agents from the FBI and DEA would be coming tomorrow to interview me. I wasn't happy about that. I had no reason to suspect the psychologist, but I didn't trust anyone from the federal government. I had no idea who was involved with Operation Juggernaut. They'd probably want to silence me if they found out what I knew. I thought about all the ways they could kill me without Officer Coffin suspecting anything. They'd probably sneak off with my confession to destroy the evidence.

I forgot to ask about calling my family. I'll do that in the morning. No reason to wake them up now.

My cell is about half the size of my college dorm room and roughly ten degrees warmer than the interrogation room. It smells clean enough, and the blanket on the small bed is scratchy but warm. Unlike my dorm room, the cell has a toilet and a sink, which are a real bonus. In college, I would have gladly traded some space for a private toilet in my room. I could see myself living comfortably like this if they didn't assign me a roommate, as they did in college. I was even starting to like the orange scrubs I'd been wearing. If I ever get out, I'd like a pair like these.

They turned off the overhead light and left me alone. There's still enough light from the hallway for me to work. Sleep is the last thing on my mind. I've wrapped the blanket around me and am determined to capture every detail before they fade from my memory. I'm on a mission; I must get my account of the events down on paper before it fades into oblivion.

After Ian's big revelation, I was driving home when Deepa called me. She started rambling about her new boyfriend, Rajeev, and all the things they'd been doing together: going to movies, eating at some fancy new restaurant downtown, and other nonsense I didn't care about. She kept saying how smart and funny he was. I had more important things on my mind, which is what I finally told her. She didn't take that well. She got all huffy and hung up on me. You'd think I'd care that my sister hated me, but I also had things to figure out and didn't have time to sort out her life.

One thing was clear: Ian had me trapped. I was driving his fancy car, had the job he gave me, and a girlfriend he'd practically handed me. Now, he had me hooked and was pulling me in.

The image of the bearded man kept replaying in my mind. I saw him thrashing and screaming. He looked at the camera for a moment, staring directly at me, pleading for my help. What would it be like to be in the switch without any way to escape? It was a punishment only the devil himself could have devised.

I gripped the steering wheel. The windows were wide open, and the wind whipped against my face. Whenever I felt trapped, I would remember being twelve years old. It was a hot summer day. I ran a race against some boys my age, but I lost at the last second. Afterwards, we wandered around the picnic tables, eyeing all the food. The aunties in their silk saris laid out flimsy paper plates and prepared the dishes. One auntie shooed us away. "We'll call you when it's ready."

The other boys ran off, but I sat at the end of the bench next to the desserts. I lifted the foil off the one closest to me. A wave of steam rose from the apple pie. I'm not sure who brought it. Bringing an apple pie to an Indian picnic is a bit unusual, but I wasn't complaining.

Soon, the smell of cinnamon filled the air. The foil stuck to the syrup as I carefully lifted it from the steaming pie. I broke off a piece of crust and ate it; it was delicious. I pressed my thumb into the scalding pie and licked my wounded finger. It was the best thing I'd ever tasted.

That's when it happened. A fly landed in the hole I'd made in the pie. It started buzzing frantically, trapped in the sweet syrup. I didn't know what to do, so I panicked and covered the pie back up. I could hear the fly beating its wings uselessly against the foil. For a moment, I thought about trying to free the fly from its prison, but fear took over, and I ran away. I never forgot that poor fly struggling for dear life. I thought of him each time I was confined to a mental hospital with suicidal thoughts. Now, I was trapped again. I was mired in a sticky mess with no hope of escape.

I'd gone nearly two months without a migraine, which was very unusual for me. However, the glare of oncoming headlights was bothering me, and I felt a major one coming on. I took deep breaths and tried to will it away. That sometimes worked, but mostly it didn't. By the time I reached home, I felt a little better. The headache never appeared, and the face of the bearded man slowly merged into the thousands of images of similar men that the military hunted down and killed every day.

"He deserved what he got," I said firmly. "We're at war, after all, and he's the enemy. He'd do the same to us if he had the chance."

I found Sanquel in the living room, reading by the dim light. He looked up from his book. "Are you hungry, my friend? The refrigerator is full. You should get something. I'll come join you."

"I'm okay," I said and headed for the staircase. Sanquel must have noticed that I looked like shit. He set his book on the coffee table and waved me over. "Come chat with me, my friend." I didn't object. I went over, sat on the couch opposite him, and stared at my hands.

"I'm afraid I've been a poor housemate," Sanquel said. "My work has kept me away."

I didn't respond.

"So, tell me about your work." Sanquel leaned in to listen. "How are things at Harvester?"

"Fine," I said, avoiding his gaze.

"Are they treating you well?"

I didn't answer. Finally, I asked, "What do you think about torture?"

Sanquel leaned back and chuckled. "Is it that bad, my friend?"

"No," I replied sharply. "I mean, what do you think about using torture to get information from people—terrorists, criminals, bad folks? Do you think it's ever morally justified?"

Sanquel looked at me, surprised. We sat in an uncomfortable silence for a long time.

"Do you mean beatings, electrical shocks, waterboarding, sexual assault, that sort of thing?"

"Yes," I whispered.

Sanquel leaned back and took a deep breath. "It happens far too often," he said. "Despite everything we've learned about how ineffective torture is, people still feel the need to inflict this kind of cruelty."

"Maybe that's true," I said, "but most Americans say they support torturing enemy combatants if it can prevent a terrorist attack."

Sanquel nodded. "The reason they think that is because they don't understand that once torture becomes an accepted practice, it rarely stays limited to extreme cases. When your uncle is tortured because he refuses to give up his land to the government, only then will the reality of torture hit home. Of course, by then, it will be too late."

"Some people don't like their uncles," I replied.

Sanquel didn't laugh or even smile at that. "Why is this on your mind?" he asked.

"It's been in the news," I said. "I've been thinking that this whole ugly business could be avoided if there were a reliable way to extract actionable intelligence without resorting to such barbaric practices."

"What way would that be?"

I shrugged. "Like, for example, a truth serum."

"Sodium thiopental," Sanquel replied. "It's been used. I know people who had it forcibly administered to them because of their dangerous ideas."

"Let's say, hypothetically, there was something that worked."

He stayed quiet, probably wondering if I was serious or if this was some kind of sick joke.

Finally, he said, "That would be much worse."

"Worse?" I looked up at him. "Worse than torture?"

Sanquel folded his hands. "It's difficult to understand the suffering people go through, the terrible conditions that arise during war, in refugee camps, and in prisons worldwide. It seems so pointless that we allow such conditions to exist. Yet, there is a reason behind causing this misery. From the perpetrator's perspective, they believe it's the only way to break the human spirit. It's easy to inflict extreme pain, kill, or make the victim watch loved ones be tortured or murdered. But even if these acts are done to us, we still have the power to resist. If that ability is taken away and penetrating the mind becomes as easy as penetrating the body, then what would remain of our humanity? If that happened, life would become one endless horror for the powerless in this world."

I didn't know what to say, so I simply nodded.

Sanquel let out a hearty laugh. "You, my friend, have been working way too hard. My partner and I will take you out for dinner sometime soon. Our treat."

I tried to smile but couldn't.

Twenty-Two

That night, I couldn't sleep. This meant no Desert Prince, no Monkey, and no training. I was hoping to lose myself in some physical exertion. I imagined the Desert Prince saying in his sonorous voice, "There is no problem so great that cannot be solved with the edge of a sharpened blade." Monkey would nod in agreement. Monkey always nodded at everything the Desert Prince said. He was such a brown-noser. An honorable battle could have settled any dispute back in the day, but that wasn't true today. The world is much more complicated than that.

I finally went downstairs and collapsed onto the couch, which was more comfortable than my bed. I wanted to get some sleep there. When I landed on the sofa, a cloud of dog fur fluttered around. Chloe struggled in the summer heat, scratching her thick coat nonstop. That evening, Chloe lay on the rug under the coffee table, scratching her ear. She ambled over to lick my face. I rubbed her head, and she closed her eyes happily. She had the softest fur.

Two minutes before midnight, I woke up. I always feel disoriented when I wake up in an unfamiliar place. The bay window creaked with every gust of wind. Chloe lay on her side, asleep on the rug, yelping now and then.

I crept into the kitchen, grabbed a bottle of water from the fridge, and drank it all in one gulp. My heart was pounding. I put on a jacket and headed out the door. I needed to keep moving. I felt a mix of excitement and panic. The events of that night would shape the rest of my life. When you think about it, every day has the potential to be that decisive moment of no return. Every choice, no matter how small, guides us along the path. And sometimes, as we stroll along thinking everything is fine, we accidentally step on a venomous snake hiding in the grass.

It was a surprisingly mild night. A gentle breeze had swept away the day's heat. I tried to imagine what Ian was doing at that very moment. He had told me that his deck overlooked a lake. I pictured him sitting there, a cigarette in his hand, gazing at the dark water. Was he thinking about me, wondering what I was going to do? Maybe, just like me, he was searching the night sky for guidance and reassurance. We were both surrendering to the murky darkness.

Ian said they bombed the terrorist camp under cover of darkness. That sounded right. Night was full of crackling sounds, mostly harmless but some deadly. Silent figures moved through the shadows, waiting to strike. The night was both dangerous and beautiful. That night, I decided to surrender to it and see what happened.

I walked the quiet streets, my mind racing. What could it mean if such terrible people were forced to reveal their secrets? These are the people who murder the innocent. Why should I feel guilty about my part in Harvester's scheme? Besides, I only played a minor role, helping prevent the perpetrator from falling into the switch. I was making the interrogation more humane.

Furthermore, scientists should not be blamed for how their discoveries are used. How could I be responsible for Operation Juggernaut? This project started long before I joined Harvester and would continue even if I left.

But all of this was irrelevant. Here was the core of the problem: I'd spent my entire life gaining skills and knowledge, but for what purpose? All my skills and knowledge would be useless if I didn't apply them. Harvester enabled me to do what I did best. I couldn't let all my effort go to waste. Under Ian's protection, I was confident I would become a leading figure in DMTA research. I would have the resources and support needed to develop the drug that could change the world, freeing people from overwhelming anxiety, depression, and everyday monotony. Everyone would know my name. And all my enemies—Cohen, Cranston, McGregor, and now Jason Chen—would regret the day they betrayed me.

But thinking about Sophie finally convinced me and solidified my decision. If I followed Ian's plan, Sophie would love and admire me. I'd have enough money to buy her a beautiful gallery to display her art. She wouldn't have to beg tiny venues to show her work. She wouldn't have to scrape by anymore. She could focus all her energy on creating great art.

Sophie would be happy and successful. I might not have all of Ravi Reddy's connections, but I would spend every dollar I earned to make her famous. She would become an object of admiration, and even my parents would have to admit that.

The sky was clear, and Arcturus, Spica, Deneb, Vega, and Altair shone brightly that night. I listened to the wind, hoping to hear some reassurance that I was on the right path. A half-moon lit the streets as the stars twinkled in the clear sky.

It was almost dawn when I arrived home. The eastern sky was crimson. Cars rumbled past. A warm breeze started to blow. I looked at my reflection in a muddy puddle to see if I appeared to be headed in the right direction. I watched the sky brighten and imagined a ladder reaching beyond the clouds. I told myself, "I'll climb that ladder. Even if it kills me, I'll climb it as high as it goes." I had made up my mind.

When I got home, Sanquel was watching a soccer match on his tablet. I was surprised to see him there; he usually leaves for work early in the morning. He said he liked working when no one was around because it gave him space to think. When he saw me, he paused the game. "I'm glad you're home. Are you okay, my friend?"

I was tired and hungry, but I needed to sleep, even if just for an hour. I had been out all night and looked terrible. I grabbed the back of a chair for support.

"I'm fine," I said, forcing a smile.

"I've been thinking about what we discussed yesterday," Sanquel said. "Do you have a moment to talk now?"

I nodded and sat down.

Sanquel took a deep breath and began. "Humans are very good at reading others' thoughts through their actions, tone of voice, and facial expressions. Some of my work involves teaching robots to do the same. But this is very different from directly accessing another person's thoughts."

What was he even saying? I just wanted to close my eyes and sleep.

Sanquel looked at me with curiosity. "Are you sure you're alright?"

"Just fine," I snapped.

Sanquel nodded and continued, "I don't want to hold you up, but after thinking about what we discussed yesterday, I realized there's a serious problem with the 'truth serum' you suggested. The system we're using to control our robot can store information, mainly about how to perform tasks. It uses a learning approach to improve over time. However, I've found that people often wrongly believe there can be no mistakes when accessing information in a nonhuman system. We discovered many errors when we tried to get the robot to explain why it did a certain task one way or another."

I nodded, doing my best to understand what he was saying.

He kept going on. "In other words, it struggles with the difficult task of self-reflection, just like humans tend to do. AI has become increasingly advanced, approaching human intelligence. And in doing so, it has encountered some of the same problems we have in understanding ourselves and our motivations. The reliability of the information the robot provides, meaning its ability to remember and transmit information accurately, depends on both the context in which the AI is asked the question and its internal state at that moment."

"So, are you telling me that your robot has a 'state of mind?'" I asked. *Is that his point?*

That is an excellent way of putting it," he said. "The robot's state of mind is influenced by its external environment and internal conditions.

The robot's response depends on the balance of those factors. The outcome is not always predictable or even consistent."

"What is your point?" I muttered.

"This is my point," he said. "The human mind is far more complex than a robot's. We should expect even greater variation and unpredictability. For your hypothetical terrorist, if his brain's state is altered or he's under intense stress, I doubt any information extracted would be reliable. Instead, you'll have better luck understanding his psychology, wants, and needs, and encouraging him to reveal the information voluntarily. That would be much more effective."

I shrugged. "At this point, it's all just theoretical." I knew it wasn't, but I needed to get away from this high-minded conversation. I understood Sanquel was trying to make an important point that I should consider carefully. Yet whatever he said vanished from my mind as soon as my head hit the pillow. It was Friday. I called in sick and didn't go to the poker game that night.

Twenty-Three

Sunday arrived, and Sophie suggested we prepare a meal for the professor at my house. She hadn't been there before and wanted to see it. I told her it wasn't much to look at, just a rundown old house, but she insisted. Later that afternoon, the two of us went to the front door with grocery bags in hand and saw movement inside the living room. Through the warped and tinted glass, a figure waved at us. I assumed it was Sanquel. I stepped inside and set down a few grocery bags. Ian was sitting on the couch.

"Hello there." Ian stood up. "Sorry to drop by unannounced."

Sophie hesitated at the door, watching him with suspicion.

I asked, "Did my housemate let you in?"

"Doors open for me wherever I go." Ian grinned mischievously and stepped forward to greet us. He was dressed perfectly, as always, in a black suit and red tie. His blond hair was slicked back and shiny.

"My name is Ian Blair," he said, shaking Sophie's limp hand. "We met briefly once at the opening of the art gallery at Harvester. You're Morey's daughter."

"I am." She studied his face. "You look very familiar. Maybe we met somewhere else also?"

Ian laughed. "Many people say that I look familiar. I look like a generic clip-art white boy. That's why people always think they recognize me."

Sophie didn't reply. Her eyes darted between the grocery bags and Ian's face.

"In the short time we've known each other, Narin and I have become close friends." Ian looked at me, then at Sophie. "Perhaps Narin has told you about me?"

She didn't reply to that either. "I think I'll put the groceries away," she said, hauling the bags into the kitchen. I offered Ian a drink, but he declined. Instead, he followed Sophie into the kitchen and leaned casually against the doorframe. Sophie looked at him with an unmistakable look of distaste and discomfort. I shot her an apologetic look from behind Ian. He sat down at the kitchen table.

"So, what do you do at Harvester?" Sophie asked.

"I run their legal department," Ian said, folding his arms. "I am a lawyer."

She nodded blankly, as if she were momentarily dazed.

"But I can tell from your expression that you don't like lawyers," Ian added, flashing his most charming smile.

"I neither like nor dislike them," she said as she emptied the grocery bags.

"The problem is that ninety-eight percent of lawyers make the rest of us look bad."

She offered a smile. I sat on a rickety chair next to Ian. He clasped his hands behind his head.

"But you are an artist," he said, "and I love artists, or at least appreciate them." He scratched his nose and cheek.

I could tell Sophie was trying to avoid looking at him. She kept her hands busy with the groceries.

Ian continued, "I own one of your paintings."

This made Sophie stop what she was doing, but she didn't turn around.

"It hangs behind my desk at Harvester. Narin has seen it. It's the one with five mangos stacked one on top of the other." Ian glanced at me for confirmation.

"I remember the piece," she said. Despite herself, she turned to look at him again: his prominent dimpled chin, blond hair, pale complexion, and sky-blue eyes.

"It's very nice," I added, trying to ease the tension.

"I thought it would be a good investment," Ian said, his lips twisting into a half-smile. He sat back and looked her over. "I couldn't help but notice that you're limping. I assume that's from the accident?"

Her face flushed, but she stayed silent.

"I knew Ravi Reddy," Ian said. "I was very upset to hear about the terrible accident. It was a great loss. I'm just glad you made it out alive." His expression was earnest and sincere.

Ian opened his mouth to speak again, but Sophie interrupted. "I'm sorry. Could you tell me your name again?" she asked. "I don't have a good memory for names."

Ian repeated his name and kept staring at her. She pulled out her phone and distracted herself with it. I had never seen her so agitated before. I needed to do something to help Sophie get out of this awkward situation. I stood up and moved between them, with Sophie at the counter and Ian sitting at the kitchen table. "You didn't mention why you're here, Ian," I said.

I tried to block her view of him and his view of her. I had a strange feeling that she was studying him, this time more discreetly, with quick, sharp, strategic glances.

"I don't want to bore you now with talk about work, especially since you have company," Ian said, craning his neck to look at Sophie again. "I just wanted to see if you had questions about the little project we discussed the other day. You weren't in on Friday, so I thought we could talk today. But we can discuss all this in the office on Monday." Ian stood to leave.

"I've thought about it, and I'm fully on board," I said as cheerfully as I could. "Wherever you go, I'll follow. I'm your man."

Ian stood up, and I reached out for a handshake, but instead, he gave me a warm hug and patted my back. "That makes me very happy," he said.

"I need to be leaving now. It's very nice to see you again, Sophie. Have a great afternoon." I saw Ian out.

I shook my head. *How the hell did he get into my house?*

When I came back to the kitchen, she was busy at the counter with the stew ingredients, her back turned to me. "So, what's up with the creepy guy?" I couldn't see her face, but her tone was cold.

"I'm sorry about that," I said. But her question was a good one. Ian had known Sophie would be with me that afternoon; he seemed to know all her movements and perhaps mine. What was the motivation for his visit today? And how had he gotten into my house?

We cooked the bean stew in Sanquel's slow cooker. I read the instructions on my phone while she added the ingredients to the pot. Soon, steam started to rise around the rim, and a spicy aroma filled the room.

Sophie sat at the kitchen table. "I've seen him before," she said. She rubbed her forehead and tapped her foot.

"You met him at the gallery opening at Harvester," I replied.

"No," she said, grimacing. "He looks very familiar."

"Maybe Ravi introduced you to him," I suggested. "They were friends." *Or maybe he follows you around and watches everything you do.* I didn't say that out loud, but that's what I was thinking.

She remained silent for a long time. Finally, she said, "I'm certain Ravi never mentioned him to me."

I pictured Ian hiding in the bushes, waiting for Sophie to head out of her house. "He's a strange guy," I said. "But, in his defense, he's gone out of his way to help me."

"Why do you think that is, Narin?" she asked sharply. Her cheeks glowed a fiery red. What she was really asking was: 'What kind of naive idiot are you?'

I realized I was in dangerous waters. I might say something foolish that could threaten our budding relationship.

"I don't know," I replied lamely.

"What did you mean by saying you'd 'follow him wherever he leads?' That's a strange thing to say to someone you work with."

I raised my hands as a sign of surrender. "All I'm saying is that he's been very generous with the job, car, and everything else. I think I owe him my loyalty."

"He has his reasons, Narin," she hissed. "People like that always have their reasons. I wouldn't trust him if I were you."

"Well, I think I know at least one of his motives," I said, stirring the soup. "For some reason, he wants to show my dad that he's got his life together." I told her about how my dad had been Ian's undergraduate faculty advisor. "Somehow, Ian thinks that if he helps me, he'll repay my dad a favor. That seems to be important to him. I have no idea why. All I know is that I've allowed myself to become a stand-in for my old man these past few weeks. Honestly, I'm using Ian just as much as he's using me."

I wanted to tell her about the Type II receptor blocker and how I developed it myself, how I brought it to Harvester, and how, with Morey's help, we would soon create a drug cocktail that would let people use DMTA without worrying about hallucinations, addiction, and, most importantly, the switch. I wanted to tell her that this redesigned DMTA pharmaceutical would be the miracle drug of the century, treating depression, anxiety, and dementia, and improving cognition and creativity in ordinary people. I desperately wanted to impress her, but I realized this wasn't the right moment. I had to wait until I had something concrete to show her so she wouldn't think I was just making shit up.

But stupidly, I also felt the need to defend Ian. "I don't know, Sophie," I said. "Ian's opened my eyes. He's a guy who knows what he wants and is willing to do whatever it takes to get it. I have to say, there's something refreshing about that. I could probably learn a thing or two from him."

"What does he want?"

I should have seen that coming. She watched every move and facial expression I made as I tried to think of something to say.

"He wants what everyone wants: to be successful," I said. "And for some reason, he wants me to be successful too."

"And what do you think about that?"

"I'm fortunate to have someone like him on my side advocating for me," I said. That's how I felt at that moment.

"But something still bothers you," she said.

Yes, something was bothering me. I was part of a highly questionable human experiment.

"No. I've got everything figured out," I said. "I have a job where I can make good money and succeed at what I'm good at. If Ian isn't just messing with me, then he'll fast-track me up the corporate ladder. And for you, I can keep an eye on Morey and make sure he doesn't get into trouble. And finally... finally, I have supervisors who won't take credit for all my hard work. Yeah, I'm good. No conflicts here."

Sophie jumped up and hugged me tighter than ever before. "Be careful," she whispered in my ear. "I know from Morey that bad things happen at Harvester. Plenty of people wouldn't think twice about stabbing you in the back."

"Don't worry." I let out an uncomfortable laugh. "As long as Ian's got my back, I've got nothing to worry about."

She held me even tighter, to the point where it hurt. "You're not a stand-in for anyone, ever," she said. Her voice was thick with emotion, and I thought she might cry. "You're your own person. You should understand that."

What did she mean by that?

"Well, okay," I said. "Thanks. I appreciate that."

She let me go, took a deep breath, and sat back down. "Oh, does Maru Chandra still work at Harvester?" She wrinkled her nose when she said his name.

"Yes, unfortunately, he does." I looked at her with curiosity. "How do you know him?"

"Does he still scowl at everyone?"

"He does."

"When we were kids, Morey used to take me to their house."

"Oh yeah," I said. "I'd forgotten Morey and Sanjay Chandra were friends."

"One time, when I was a kid, Maru shot a cap gun right next to my ear." She held up her hand and rubbed her ear. "I couldn't hear for a week."

"He's an asshole," I said. "Ian can't stand him. No one can stand him."

Sophie laughed. "He asked me to his little prep school prom one time. I was a freshman, and he was a senior, but he still asked me out."

"Did you go?"

"What do you think?" she scowled. "Believe it or not, I do have some standards." She clenched her fists and cracked her knuckles. "I told him I'd never go out with a loser like him. I heard he ended up going to business school. Having a daddy willing to pay for all that must be nice."

I looked over the recipe. "Beans, garlic, onions, carrots, celery, olive oil, bay leaf, thyme, paprika, salt, and cracked black pepper. It looks like we're ready. Now we have three hours to kill. What do you want to do?"

"I want to see your room."

Twenty-Four

It happened quickly. We had been sitting on the bed talking for over two hours when she leaned over and kissed me. She had a mischievous look in her eyes, so I kissed her back. We sat on the bed, kissing, when my fingers started fumbling with the buttons on her blouse. When I realized she wasn't going to stop me, I panicked.

"Are you sure you want to do this?" I whispered. I was so nervous that my hands trembled. She gently stroked and kissed my face.

I was so stupid. I should have been better prepared. I should have picked up some condoms at the store. Who cares if it's embarrassing to buy them? It would have been worth it. I thought about digging through Sanquel's room to see if he had any, but then Sophie reached down and grabbed her bag. She pulled out a condom, smiled, and handed it to me. She was ready. She knew what was coming. I was so surprised that I was speechless. I didn't think it would be so easy.

We hurriedly undressed, feeling like we had to do it before the moment slipped away. Soon, I was on top of her, and she welcomed me in. She was so incredibly warm. But something felt off. She looked distracted, lost in thought. I could see from her eyes and movements that she was trying to gather herself and return from some distant place her mind had wandered to. It seemed like a jumble of buried feelings was surfacing, and she was struggling to push them back down. What was she thinking about? Who was she thinking about?

But I knew deep down she was thinking of Ravi, and I felt a surge of jealous rage rising inside me. I should have stopped then, but my heart was pounding. Instead of stopping, I pushed harder, trying to erase her memories of him. I was so angry and desperate that I didn't know what I was doing. I wanted her to see that I was better than her old Red Skyer boyfriend. When Sophie cried out in pain, I recoiled, devastated to see her face twist in agony. Why would I try to hurt the person I loved most?

"Are you okay?" I was so upset that I almost broke down in tears.

"Let's stop for now," she said softly.

I quickly pulled back and rolled away from her.

If she was angry with me, she didn't show it. We lay in bed for a long time, with my arms wrapped around her. Then something caught her eye: Tara's star picture hanging on the wall. She stood up and went to it, tilting her head as she studied it closely. I stayed in bed and watched her stand there naked, her legs crossed. Her curious hand moved from star to star.

The timer went off in the kitchen, startling both of us. We dressed quickly and went downstairs. She took her time going down the stairs while I stayed nearby, ready to catch her if she lost her footing.

The stew was ready. She scooped up a spoonful and slurped it down.

"Pretty good," she chirped.

As we headed to the car with the stew, I noticed she was limping worse than before. My heart was still pounding, now with fear that I had made her injury worse. I was terrified I'd caused some permanent damage.

When we arrived at the professor's house, he greeted us in his usual outfit. I realized I had never seen him wear anything different. Both he and his house always smelled clean, with a faint floral scent hanging in the air. As we set the table, I stayed close to Sophie, checking on her every few minutes. She grew tired of this and gave me several tasks to do in the kitchen. She needed some space, and that was okay.

The stew was a big hit. To Sophie's surprise and delight, the professor finished two bowls. After dinner, we went out onto his porch. It was a warm evening. The professor gently swung in his wooden porch swing, which creaked on rusty chains. Insects bounced off the glowing, dirty globe overhead. The professor sipped his sherry. We sat on the porch steps, holding hands in the shadows. Everything was quiet and peaceful.

The professor spoke. "Your father and I have been communicating by email, Narin."

"Is that so?" I replied. I realized I hadn't checked my personal email in over a week.

"He tells me that you've been studying the compound savycha, which has been used in shamanic rituals in South America and has now gained notoriety because of this unfortunate business with Red Sky."

Sophie let out a small gasp as she uncrossed her legs and let go of my hand.

"That's been the focus of my research for the last few years: understanding its chemistry and how it affects the brain," I replied. I watched Sophie for any signs of discomfort, but her expression remained inscrutable.

The professor nodded. "I visited a small mountain village in Bolivia many years ago. I spent only a brief time there, as it was just one stop on a longer journey. I went there because I had questioned a German scholar's earlier work on the myths and stories of these people along the Altiplano, or the Andean Plateau. He seemed to have rather romantic notions and a fanciful imagination. My time there was brief, but I did have the privilege of experiencing savycha myself."

"You used the drug?" I asked, surprised by this revelation.

The professor chuckled. "The people of the Altiplano do not consider it a drug as we would use the word," he said. "It is not part of their usual religious practices, as some have suggested. But in rare cases, such as during war, plague, or funerals—I was there when the village matriarch had just passed away—during these rare moments of great distress, when guidance from and connection with the eternal are sought, then savycha is used for that purpose."

Stars started to appear as the sky darkened.

"It was the strangest sensation," the professor reminisced. "The whole world pulsed like a crimson heart."

The three of us sat quietly on the porch. The old professor rocked gently as the start of the cicadas' nightly song hummed in the darkness. I could have happily sat with the professor and Sophie like this forever. I looked at Sophie in the dim light. She wasn't upset and probably thought I was being too forceful because of my inexperience. I hated to think she suspected I harbored hostile thoughts about her. I decided I would never say or do anything to hurt her again.

The silence was broken when Sophie spoke. "Professor, I read that you were accused of being an atheist after your book came out. I find that very strange. It seems like an incorrect interpretation of what you were saying."

The professor nodded. "I've been accused of much worse." He took a sip of his drink. "Much, much worse," he said, gazing into the darkness.

Twenty-Five

JoAnn and Ahmed's complete disregard for keeping me informed, whether intentional or not, was growing worse. I felt I had to confront them. I found JoAnn sitting in her office, reading a journal. I walked in and closed the door. She looked up.

"Hi, Narin," she ventured. "What's up?"

I sat down and looked at her but said nothing. "Is there something wrong?" she asked testily.

"We're a team, aren't we?"

She didn't respond.

"You and Ahmed are shutting me out. You haven't been telling me everything you're doing with H138 or the Type I receptor," I said flatly.

She scowled. "It's all on the Harvester project site." She considered for a moment. "Ahmed and I are not the ones who keep a separate lab book."

"Still, I need to know what you're doing," I said.

She flushed and replied, "Well, if you didn't spend so much time upstairs hobnobbing with Ian, you might better understand what's going on down here."

"I hardly ever go up there," I shot back.

"You know the only reason Morey hired you was because you had a Type II receptor blocker," she said with venom. I saw in her eyes how much she resented having me there.

"So, I'm useless now?"

"No," she yelled, "I'm saying that your part of the project involves the Type II receptor. We don't get into your business."

"So now your work is your own business, and I should just butt out?" I was losing control, but I didn't care. "Is that what Morey wants?"

She slammed her hands on her desk and said, "Well, if you don't trust me and Ahmed, then maybe Harvester isn't the right place for you."

I nodded. "I just don't want to be shut out." I was surprised at how assertive I sounded. And I liked this new assertiveness. I wasn't sure if it was Ian's influence or the Desert Prince.

I got up and went to the door. But Joann wasn't finished yet. "By the way," she said, "Morey let it slip that you came on with a higher salary than mine, which is really unfair since I've been working my ass off on this project for a year now. But I understand how things work. Female scientists aren't valued as much as male scientists. That's just a fact."

"You're right," I replied. "That is unfair." I had to admit. She did have a point.

The elevator opened on the ninth floor, and I searched my pocket for the key Ian had given me to the room next to his. The heavy black door swung open, revealing an office similar to Ian's. The plush crimson carpet covered the entire floor. The desk was large and imposing. I saw my reflection in the gray marble desktop. The computer had dual high-definition monitors. A flat-screen television was mounted on the wall. There was a refrigerator, a freezer, and even a reasonably comfortable folding bed. Next to the wall dividing this office from Ian's, there was a small cabinet filled with a variety of tumblers and shot glasses. I looked but couldn't find any liquor stored inside. Ian had said this would be a great place for me to relax and unwind. He was right about that.

I sat in the ergonomic black leather office chair and rested my feet on the shiny desktop. I thought about what it would be like to be one of the executives, maybe the Vice President of Drug Development, Maru's job. This is where I'd spend most of my time. I knew I could do that job, and I could do it a hundred times better than Maru. I could lead Harvester into the future and help develop the pharmaceuticals that could change the world.

Then I heard Ian's voice. He was talking to Maru next door. I couldn't catch all the details, so I grabbed one of the tumblers and pressed it against the wall. To my surprise, it worked. I could hear their conversation as if they were in my room.

Maru said that Morey didn't respect him and would bypass him to talk to his father, Sanjay Chandra, whenever they disagreed. Ian responded that since Sanjay and Morey had been friends for more than twenty years, it was only natural for Morey to go directly to Sanjay. Maru should accept this.

My cell phone rang, and the conversation in Ian's office abruptly stopped.

"I wanted to let you know that they brought the Type II receptor blocker to the lab," Ahmed said. "It's labeled H140. The specs show that it binds to and blocks the Type II receptor with high affinity. You were right. If it doesn't activate the receptor and blocks DMTA binding, it's exactly what we've been searching for."

"It doesn't activate the Type II receptor," I whispered. "It just blocks the binding of DMTA. Does Morey know that we've got it?"

"Morey knows," Ahmed replied. "He's already given us our marching orders. He wants a report on the effect of H140 on DMTA binding at both Type I and Type II receptors, with and without H138 present. I'll get everything ready. You're the expert on this, so I want to make sure everything looks good to you before we get started."

JoAnn probably told Ahmed about our earlier conversation. This was his way of calming me and making peace.

"I'll be right there," I said.

I exited and saw Maru coming out of Ian's office. He waved awkwardly. I pressed the elevator button several times, eager to leave. That's when I noticed Maru walking over.

"So, you've been with Harvester for a couple of weeks now," Maru said. He had such a squeaky, high-pitched voice that I almost laughed.

"Almost two months," I replied without looking at him. I saw his reflection in the elevator door.

"Have you been able to make any friends?" he asked.

I stared at his reflection. He looked at me through his oversized glasses, his eyes full of contempt.

"I've managed," I said.

"Come by my office sometime," he said. "We should talk. I think we have a lot in common, you and me. We should hang out together sometime." Maru smiled awkwardly.

To my relief, the elevator arrived, and I stepped inside.

"And try to stay out of trouble," Maru said as he walked down the hall.

"You do the same," I yelled as the elevator doors closed.

That evening, when I left work, I looked up at the full moon and thought about how the sun and moon always rise and set in the sky. For some reason, this made me think of Sisyphus and his endless punishment of pushing a boulder up a steep hill each day, only to have it roll back down at the end of the day, repeating the cycle forever. I wasn't going to be like that. I wouldn't let all my hard work go to waste. My efforts would mean something. I was in the game to win, even if it meant risking everything.

Twenty-Six

The following week was the busiest of my life. Morey, JoAnn, Ahmed, and I worked from morning until midnight, trying every possible version of experiments.

My Type II receptor blocker worked perfectly, and the presence of H138 did not reduce its effectiveness. This showed that DMTA, H138, and H140 (my Type II receptor blocker) could be used together for Operation Juggernaut. H138 would enhance DMTA's effect and induce a trance-like state, making the subject highly receptive to suggestion and questioning. H140 would prevent the switch and the shift to violent suicidality. I wondered whether JoAnn and Ahmed understood the true purpose of this project. They probably suspected something based on our experiments, but maybe they had learned from experience that it's safer not to know the full story.

I couldn't visit Sophie that week. Instead, I would go up to the ninth floor when I knew Morey wasn't there and call her in the evenings. We talked about silly things. She'd tell me how god-awful all her paintings were turning out and how she should just get it over with and kill herself. I'd share all of Morey's antics with her, and she'd laugh and tell stories of all the crazy things he did when she was a kid.

It was on one of those evenings, while talking on the phone, that we had our first real fight. We'd had minor disagreements before, usually when I dared to criticize Goldblum's famous book. She loved Goldblum and fully embraced his philosophy of focusing on the "empty spaces" in life. She said she was practicing "listening to the silence," whatever that meant. She told me it was helping her with her anxiety issues.

I tried to explain to her that Goldblum was making all sorts of assumptions about the nature of existence without offering any proof. She said her feelings and intuition were the only evidence she needed. I didn't push the point. It wasn't a dealbreaker that she believed in all of Goldblum's New Age nonsense. It was a harmless enough philosophy, and even though she was annoyed by my critique, it didn't seem to bother her that I was a 'non-believer.'

Like most arguments, this one started with a simple question.

"So, what are you guys working on?" she asked.

Of course, I wasn't going to tell her about Operation Juggernaut. Instead, I shared what I hoped to work on soon.

"You know how people use DMTA to be smarter and more creative?" I asked. "We're creating a way for people to get all the benefits of DMTA without the risk of addiction and suicide."

She said nothing, and I wondered if she was still there. Sometimes, she would start working on something, and I could tell she wasn't mentally present by the way she responded to my questions.

"So," she started, "you're making a pill to make people smarter and more creative?"

"Well, people already use DMTA for that," I explained. "We're just making it safer for people who use the drug. It's really about creating a form of DMTA that anyone can use without worrying about their safety."

"Why?" she asked sharply.

I was surprised by the question. The answer was clear to me. Before I could reply, she continued, "Don't the rich already have enough advantages? Why give them more tools to gain even more?"

"I mean..." I wasn't exactly sure what she was getting at. "We're not preventing anyone from doing what they need to do to succeed. We're just adding a new tool for those who can—"

"New tools for those who can afford them," she interrupted. "It's unfair, Narin." She was shouting now, and I had no idea why. "Doping is unfair. It gives people an unfair advantage, and it makes the people who don't dope feel like they're suckers for not doing it also."

I was about to point out that her ex-fiancé owed all his artistic success to DMTA use, but I chose not to. I needed to diffuse the situation before it worsened. I had to avoid saying something I couldn't take back.

"Sure, it will give people an advantage," I replied cautiously. "But so do other things. For example, you went to art school, right? That gives you an advantage over people who didn't have the time or resources to do that."

She fell silent again. *Shit, that was an idiotic thing to say. I just made things worse.*

Finally, she spoke, or I should say, yelled. "Art school was fucking hard, Narin. I worked my ass off to get the skills I have. It's not like taking a pill."

I had no idea how to escape this situation. I kept going, explaining that we're still in the early stages of development and that the goal is to create a drug that everyone can take. I assured her this would benefit society. The people who use it would make significant advances that would eventually help everyone.

Thank goodness she let it go. We talked a little longer before she hung up. Usually, we'd have a playful argument about who should hang up first, each claiming we wanted to talk longer. Not this time. I could tell she was upset. Somehow, I'd hit a nerve. I wanted to call her back and apologize, but I wasn't even sure what I'd done wrong.

Under normal circumstances, I would have been angry about such a misguided critique of my life's goal. What's wrong with developing a drug to make people happier and smarter? Still, I couldn't be mad at her. I loved her too much. After all, she wasn't a scientist; she couldn't be expected to understand what I was doing.

It was during this time that I finally defeated Monkey. He would attack, and I'd maneuver from side to side, ducking and dodging to avoid his blows. I was always searching for an opening. I was growing stronger and more skilled, able to deflect most of his attacks. The Desert Prince had previously called my attacks obvious and foolish, but he stopped saying that. I used all my strength and wits to get close enough to Monkey to land a fatal blow. Each time we sparred, I was getting closer.

Then, one day, I realized something. When he attacked, he used his left forearm against my chest to block my body while he struck my chest or abdomen with his right arm. I waited for this and gave him an opening to lure him into attacking. When he did, I pushed his elbow off my chest as hard as I could before he could stab me. This knocked him off balance, and I was able to get behind him. I grabbed his head and slit his throat from behind with one clean cut. Monkey collapsed like a rag doll to the floor. A pool of red blood flowed from his neck.

The Desert Prince said nothing but nodded his approval. *I won. I had fucking won! This was the best day of my life.* There was no feeling more incredible than standing over the lifeless body of your adversary and watching his life drain into a pool of bright red blood.

I reflected on the two times I was hospitalized for suicidal thoughts. What was suicide but misplaced aggression? I had felt so powerless that I turned my hatred inward. But I would never let that happen again. It is my enemies, not me, who would feel my wrath.

I kept going to the poker games. Sometimes I won, sometimes I lost, but I mostly did it out of habit. Nothing interesting was said. One night, Ian asked how people felt about Ann wearing glasses. I didn't know who Ann was, but I later found out she was the disliked head of HR, the skinny woman who led our orientation. Chris said he thought she actually looked better when *he* didn't wear his glasses. Everyone laughed at that, including me.

On Saturday, probably late June by then, Sophie and I were heading out for dinner with Sanquel and his partner, Peter. I waited in my room for Sophie to arrive. The air was hot and sticky. She walked up the path wearing a wide-brimmed straw hat she bought at a thrift store. She climbed the crumbling steps. I watched her from my upstairs window.

She tentatively grasped the handle of the screen door. The door was open, but she was probably debating whether to go inside or ring the bell. I should have gone downstairs to greet her, but I wanted to see what would happen when she met Sanquel. I hadn't described him to her and was curious to see how she'd react.

Sanquel pushed open the screen door, and the handle slipped from her fingers.

I heard him laugh. "We are being somewhat lax with our security today," he said. "Fortunately, you don't look like you've come to burglarize us."

She extended her hand and shook his. "I'm Sophie," she said.

"My name is Sanquel," he said. "I'm glad you could join us, Sophie. I feel like I already know you. Narin speaks about you warmly and often."

I wish I could have seen her face. I'm sure she was blushing. She blushed easily, especially when meeting strangers. But the hat was in the way. I watched her rub her sandal in a wide arc across the patio. It made her look like a schoolgirl.

"Please come in," Sanquel said. "I have freshly made coffee if you'd like some."

"Thank you. I would," she said.

I could smell the coffee from upstairs. I went downstairs but sat on the staircase, listening. Sanquel filled a mug with coffee in the kitchen and offered her cream and sugar. I crept a little closer until I was in the living room. She took her mug and sat at the kitchen table with her legs crossed and her head down. I could tell she was watching him. She drank her coffee while he washed his mug in the sink.

I'd seen enough. I went into the kitchen, and she offered her cheek, which I gently kissed.

"My partner, Peter Ferrell, will join us today," Sanquel said. "We should probably leave soon."

Sophie finished her coffee and passed the mug to Sanquel.

We headed to the street where Sanquel's car was parked. "There are many wonderful restaurants in St. Louis," Sanquel said. "But the place I've chosen serves authentic Mexican food in the way I most enjoy. I hope that you'll find the cuisine agreeable."

"Sounds great," I said.

We both insisted that Sophie sit in the front passenger seat while I took the back. I watched her examine the different dials lighting up on the screen as Sanquel drove.

He noticed her staring at the screen. "These hybrid cars bombard you with real-time fuel efficiency and battery use statistics," Sanquel said.

She sat with her hands in her lap and her shoulders hunched, her head leaning forward. I knew her well enough to realize that if she thought it was appropriate, she would pull her legs up and squeeze them against her chest, resting her chin on her knees. She always preferred sitting like that, curled up in a fetal position.

"I told Sophie that your dad was a well-known artist," I said.

"Indeed," Sanquel replied. "Unfortunately, I haven't inherited any of his talent."

"He was one of the founders of the Rupture Movement," Sophie said. "At least, that's what we learned in school."

Sanquel looked at her and smiled, "If he were alive, he would reject that characterization. As often happens, a feud over trivial matters led to rivalry, hostilities, and denunciations. He made a dramatic public declaration in his later years, renouncing and condemning all his early works as expressions of capitalist decadence. But indeed, you are correct. His association with that movement was his claim to fame."

I asked, "Do you still have any of his paintings?"

Sanquel sighed, "For now, no. One of my goals is to acquire as many as I can if I ever get the financial means to do so. His most famous works are kept in private collections around the world. If my father knew his prized paintings were hanging in the homes of hedge fund managers and CEOs, he would, what is the expression? — roll over in his grave." Sanquel laughed. "But Sophie, I've heard that you're quite the artist yourself."

"She's very talented," I said.

Sophie remained silent, watching the dials on the screen shift into green or orange arcs whenever Sanquel sped up or slowed down.

He looked at me in the rearview mirror, and I gave a small nod in reply.

"There is a group of artists in Chicago whom I know well," Sanquel said. "I mention this because they hold a retreat every year in the fall, where they present their portfolios and discuss their work." Sanquel glanced at Sophie, who sat still with her head down.

He continued, "It's a relaxed kind of event. Everyone is very supportive. Some in the group want nothing to do with the business side of things, but others see it as their main goal to find ways for artists to sell their work. Anyway, it's happening in October. Both of you are welcome to come." Sanquel looked at her again. She sank further into her seat, her eyes still fixed on the display.

"If you wish, Sophie, you can present your work," he said. "You will find a warm and encouraging reception. And the two of you won't have any trouble finding a place to stay." Sanquel laughed. "It's not what you might expect, a rathole in an abandoned building. Several good friends have flats on Lakeshore Drive, which are all quite nice. I plan to visit for a few days to catch up with friends. Perhaps we can all travel together."

I was going to suggest that this could be a good chance for her to meet my family, but I held back because it was still too soon for something like that.

"I think that's a great idea," I said. Sophie nodded slightly but didn't say anything. We continued driving in silence.

"Ah, Narin, I've meant to ask," Sanquel said. "Have you finished your work on the truth serum you're developing at Harvester?"

I glanced at Sophie. She turned to face me, her brows furrowed. She mouthed the words, "Truth serum?" I shook my head. What was I supposed to say? I wasn't sure if Sanquel was joking, but honestly, this wasn't a joking matter.

Sanquel turned onto Cherokee Street. Sophie and I watched the vibrant signs of the passing buildings. We heard snippets of Spanish as teenagers playfully teased each other on the sidewalks. The street was crowded with people entering and leaving bakeries, restaurants, ice cream shops, coffee shops, yoga studios, and supermarkets. This place was so different from any other I'd seen in St. Louis that I wondered if we had been transported to another city or even another country.

"What splendid luck," Sanquel said. "We have parking right in front."

I saw someone jump off a bench and wave his arms wildly at us. We parked the car and got out.

"Peter, I hope we haven't kept you waiting too long," Sanquel said, embracing him and giving him a kiss. "This is my business partner and life partner, Peter Ferrell. And Peter, these are my friends, Narin and Sophie."

"Oh my gosh," Peter said, "look at these two, Sanquel. Not only are they gorgeous, but you can tell by looking at them that they're a power couple." He shook a finger at us. "We had better be careful with these two. We wouldn't want to mess with them." I went to shake his hand, but he embraced me and did the same for Sophie. She giggled as he squeezed her.

Peter led us to the restaurant. "I've already got a table picked out, our favorite, away from all the noise." He turned to us. "You guys will love this place. It's Sanquel's favorite; he knows his food better than anyone. Food is what keeps us together. He loves to cook, and I love to eat."

"He doesn't look like a big eater," I whispered to Sophie, but she didn't respond.

We waited as patrons shuffled in and out of the restaurant, with music spilling out every time the door opened. The contrast between Sanquel and Peter was striking; while Sanquel was dark, muscular, and had perfect teeth, Peter had red hair, pale, freckled skin, stick-thin limbs, and teeth jutting out in all directions.

We entered and were led to a back room overlooking a patio. The loud music from the front was just a muffled rumble in the back. The walls were tiled and depicted an idealized rural scene from Mexico. The ceiling had exposed brick, intersected by steel pipes and vents. Sanquel and the waiter exchanged a few words in Spanish, sharing a laugh. Sanquel said in English, "I do not doubt your team's ability, but at the present moment, Spain is up by two goals."

"Just wait," the waiter said with a smile before leaving.

"I've asked them to bring us their house special spiced sangria," Sanquel said. "It's quite delicious." I didn't want to tell Sanquel that I couldn't drink red wine; it always gave me a vicious migraine.

We looked at the menu, and Sanquel suggested food options for the group. We decided to order several dishes to share. When the sangria arrived, Sanquel poured glasses for everyone and placed the food order. I took a small sip and whispered to Sophie that she could have my drink. She didn't respond, but she pulled it closer to her.

"I know you've been curious," Peter said. "But you're too shy to ask. 'What the hell is up with that guy's teeth?' The story's pretty simple if you want to hear it. I used to get the shit beaten out of me regularly as a teenager. My teeth got busted up, and my parents refused to fix them."

"That's terrible," Sophie said. I noticed she had already finished half of her glass of sangria.

"I like it," Sanquel said, grabbing Peter's cheeks. "It gives his face character."

"Yeah, well, you're not the one who has to eat a cheeseburger with this mess," Peter replied. "My dad, when he found out about my predilections, would call me a bitch of a son. You know, instead of an SOB, I was a BOS. He thought that was just hilarious."

"And this was when Peter was only thirteen years old," Sanquel said. "But now they have reconciled, and all is well. Isn't that correct?"

Peter shrugged. "Did you have any issues with the car?"

Sanquel shook his head. "Thankfully, no."

"You might want to slow down, girl," Peter said to Sophie, "this isn't your Mama's sangria. It's more like brandy than orange juice."

Sophie set down her glass and shrank back into her chair.

"You don't drive much anymore, Sanquel," I observed. "You ride your bike everywhere."

"It is the first time I've driven in a while," Sanquel said. "And yes, everything went smoothly. Peter is very skeptical of this new hybrid technology. He has his reasons, which I am not smart enough to understand."

"If I didn't design it, then I don't trust it," Peter said. "But I always tell him he's crazy to ride his bike in this city. Bike pattern blindness is an epidemic here. They'll run you down without even noticing and go on their merry way."

"It's not like Chicago," Sanquel said.

"No, St. Louis isn't Chicago," Peter said.

"That's true," I said.

"I'm what you call a casual cyclist anyway," Peter said. "I wouldn't dare ride my bike on the streets. But even then, there are these total assholes," he grabbed his shirt, "who wear their little tight-fitting biking uniforms and have their fancy shoe clips, and they whiz by at top speed, inches away from you, passing without any warning. I just want to yell at them, 'Hey asshole, this is a family-friendly bike path in Forest Park, not the fucking Tour De France.'"

The food arrived, and we started sampling each dish. Sanquel lovingly described the feast before us. There was an incredible meat stew called birria, apparently made from goat meat marinated in chiles. We scooped up this dish with handmade tortillas, staining our mouths and fingers red with the spicy broth. It reminded me of homestyle Indian cooking. There were tacos al pastor, which I thought I was more familiar with until I was surprised by the chunks of pineapple and tender pork. The highlight was white fish ceviche served on tiny crispy corn tostadas, each of which seemed to burst with flavor and refreshment in my mouth. The food was so captivating that we hardly spoke until we were completely satisfied.

"So, how did the two of you meet?" Sophie asked.

Peter and Sanquel exchanged glances. Peter began, "He chose me to be his partner in a freshman robotics class when we were undergrads."

"That isn't exactly right," Sanquel said. "We were assigned together."

Peter looked confused. "I thought that you chose me."

Sanquel shook his head.

"I thought you told the professor you wanted to work with the funny-looking redhead with the goofy smile."

"No such thing happened," Sanquel replied. "It was purely happenstance. Would you like me to tell the story?"

"I guess since I obviously don't know what I'm talking about," Peter said, taking a large gulp of sangria.

"The short version is this," Sanquel said. "Each group had to develop a drone capable of flying in a specific geometric shape. This was in the early days of drone technology, and the best way to maneuver these devices was still being figured out. We were tasked with flying the drone in a circle, which probably seems like the simplest trajectory imaginable. Peter built the drone, and I programmed it."

"But it kept making these random errors," Peter interjected. "And Sanquel kept saying it was a design issue, not a software issue."

"We spent hours correcting each error," Sanquel continued. "But then it would make another random error on its next flight. We were quickly approaching the deadline, and our drone couldn't complete the task within the required specifications."

"So, what did you do?" I asked, glancing first at Sanquel and then at Peter.

Sanquel chuckled. "Peter had the clever idea of flying the drone a thousand times, gathering data on the errors but not fixing them. After all the tests, I created an algorithm based on averaging all the errors. Since the errors were random, they canceled each other out. And on demonstration day, our drone made a nearly perfect circle."

"The only thing I'd add is that it was closer to five thousand trials," Peter said. "And we were fortunate that it worked."

"Indeed, we were lucky," Sanquel said, leaning over to kiss Peter's cheek. "Then, in our senior year, we decided to start a robotics company. We had no idea how to do it, but we were fortunate enough to join a tech incubator with talented engineers and programmers just down the hall. We developed our prototype two years ago and received seed funding for production from an investor in St. Louis. So, we moved here, rented space in the Cortex District, and are very close to having a commercially viable product: a general-purpose robot."

"It's getting exciting now," Peter said, rubbing his hands. "Sanquel, how much can we tell them?"

Sanquel shrugged.

"I'll tell you what. I'll tell you guys everything, but if you tell anyone, I mean anyone, I'll have to kill you. Deal?" Peter reached across the table and shook hands with me and Sophie.

Peter leaned toward us conspiratorially. "Traditional robots have a computer inside to manipulate the actuators, making the robot move and perform useful tasks. But that's not how movements happen in real life. Even simple creatures are covered by information-gathering nerves. The movement itself triggers these nerves to fire, providing the brain with feedback on the movement. So, living organisms can make instantaneous, small, real-time adjustments to correct errors. Well—we're very close to having a commercially viable robot that can do just that."

Sanquel added, "So, imagine sending a robot into a burning building to rescue those trapped inside. The robot can send information back to human rescuers, but then the rescuers will have to risk their lives to perform the rescue. Ideally, you want a robot that can perform the rescue independently."

"Now, Sanquel, there is a third option," Peter said. "You can have an operator on the outside controlling the robot, but the problem is that communication can be disrupted, or perhaps the rescue is happening in outer space, where there will be a time delay between the human controller and the robot. That's why you want a self-contained system that can make decisions on the fly and doesn't require a human interface."

"So, you're creating a sentient robot," I said. I meant it as a joke, but I could tell from their expressions that they didn't find it funny. They both looked at me.

"Yes, we are," Peter said finally. "And we're calling her Christine. Since we can't have a daughter, we'll create one. Isn't that right, Sanquel? I hope she loves us as much as we love her."

"I'm not sure what sentience even means in this context," Sanquel said. "We're creating a self-contained system that can use a wide range of feedback to guide its movements when it has incomplete environmental information. And frankly, even with our limited resources, we are currently further along than anyone else."

Peter wrapped his arm around Sanquel. "That's right, honey. We kick serious ass."

Sanquel laughed.

"Is the military going to be one of your customers?" Sophie whispered. She had been so quiet, moving her food around her plate without eating, that we almost forgot she was there.

Peter and Sanquel looked at each other, and Sanquel replied, "No, Sophie, we will not be selling these to the military. A letter has been circulated among all the robotics companies in the United States, pledging not to allow the devices we create to be used for military purposes. We are just a small player now, so it was easy enough for us to sign such a pledge. However, to the credit of many larger companies, every major corporation working in this field has signed the pledge. Rest assured, there will be no autonomous killing machines in your lifetime."

Peter placed a hand on Sanquel's shoulder. "Whenever the militarization of robots comes up, I think of that quote from Isaac Asimov: 'Violence is the last refuge of the incompetent.' Sanquel and I want nothing to do with that."

Sophie and I took a walk after lunch.

"You didn't say much at lunch," I said. "But you didn't need to. Peter likes to talk."

"And I like to listen," she said, pulling her hat down to shade her eyes.

"Sanquel says that Peter is a genius."

"I'm sure they both are," Sophie replied with a hint of a smile.

My hackles rose. Was it a mistake to introduce Sophie to Sanquel? I mean, he's an attractive guy. And she was right; he probably was a genius, or at least he talked a good game. But he was gay, or maybe bisexual. Whether gay, bi, or straight, Sophie could still find him attractive. *This was ridiculous.* But all things considered, I bet I could beat Sanquel in a knife fight, just like I'd defeated Monkey. I just hoped it didn't have to come to that.

Twenty-Seven

On Sunday morning, I stayed in bed, dealing with a severe migraine, the first in over two months. I couldn't believe that just one sip of that sangria caused it. I pulled the curtains shut to block out the bright, clear day. By the afternoon, my headache had eased thanks to two doses of my rescue medication.

I dragged myself out of bed and went to Sophie's house, surprised to find the door open and her not there. I sat on her couch, ready to read my new book, *Playing Winning Poker*. That's when I noticed Sophie's sketchbook on the coffee table; she'd never left it out like that. I picked it up and flipped through the pages, shocked to see Ian's face staring back at me. She had drawn his portrait in charcoal but took the time to color his eyes with a blue pencil. A curl of wavy hair framed his handsome face, reminding me of an anime character—a dashing hero, a heartthrob. He looked like a perfect man, or rather, a man whose flaws were charming. I'd say she captured Ian's essence.

I flipped through the other pages, knowing I wouldn't find a picture of myself. I couldn't help feeling incredibly jealous. She might love me, but it wasn't because of my looks. I was just a skinny, nerdy Indian kid with nothing appealing to offer. She tolerated me because I reminded her of someone she truly loved but could no longer have. I was a fucking stand-in for a better man.

I didn't blame her, of course. Ian was so much more interesting than I was. He was smart, wealthy, talented, and charismatic. I couldn't blame her for being obsessed with him. Hell, I was obsessed with him, too. I reflected on all my failures in life, all the times girls had scorned or laughed at me. They were right to hate me. I wasn't attractive like Ian. I didn't even deserve their attention. I was as forgettable as they came.

I flipped to the next page. Now, this was interesting and unexpected.

In a Victorian garden, a young woman resembling Sophie rocked a cradle. Her head was tilted back, her eyes soft and pensive, and her lips parted as if she were singing a lullaby. Flowers bloomed around her, and large fronds cast shadows on the pavement. A butterfly delicately balanced on a branch — but wait — the cradle was empty. *Shit.* The scene couldn't have been more disturbing if blood had been splattered across the page. Why was Sophie rocking an empty cradle? Why was she singing to a nonexistent baby? Was this an insane asylum? Did the baby die? Or had the infant been taken away?

When Sophie returned an hour later, I was sunk into the couch with my feet on the coffee table, absorbed in my book about poker. I had carefully placed the sketchbook back where she had left it. I was still sore from what I'd seen in her sketchbook. Maybe she secretly loved Ian. Maybe she still secretly loved Ravi. But none of that mattered. She was with me now, and I was with her. I needed to make the most of the situation.

"Hey, you," she said. "How'd you get in?"

"Doors open for me wherever I go," I replied, laughing. "I just pushed on it, and it opened. I thought it was a little strange you'd left it unlocked."

Sophie examined the door. "How hard did you push?"

"Not hard." I sat up and moved closer.

She ran her finger along a crack on the wooden frame, which was damaged near the lock.

"I wonder," she said. "I wonder if someone tried to break in?"

I examined the lock. "It's broken. It won't latch." My first thought was that Ian did this — that he had broken into her house. But that didn't seem like him. He's a creep, but this wasn't his style. If he wanted to get into Sophie's house, he would find a more sophisticated way to do it.

Sophie sighed. "That's just great. There goes my monthly savings."

I closed the door then opened it again. "I'll tell you what. I'll pay to get it fixed."

She glared at me. "Why?" she said testily, "It's not your fault. It's my own fault I live in a neighborhood with porch pirates and petty criminals." She stormed through the house, searching for missing valuables or signs of a break-in. She was furious. I just hoped she wasn't furious at me for offering to help her.

I asked, "Were you at the UPS store?"

"I was," she said, rummaging through a bag on the dining table. "I sold three pieces in the last two days and made five hundred dollars." She tried to smile. "I'm not famous yet, but people are starting to notice."

"We should celebrate," I said, pulling her close and kissing her.

She pulled away; she wasn't in the mood for that.

"I'm still making more from the knick-knack stuff, but if this keeps up, I'll focus on painting. I'm teaching a class next spring, so I'll have more breathing room then." She placed her hand on her chin and furrowed her brow. "The wooden masks sold quickly last year. Maybe I should make a more elaborate set this time."

She removed the supplies from her bag and set them on the table.

"Did you carve the masks?"

"I carved and painted them," she said. "Yeah. Maybe I'll do that."

"Do you do a lot of wood carving? Like animals and stuff?" I was thinking about the wooden monkey in my dad's study. Maybe she could make one like that for me.

"I used to." She looked at her supplies scattered across the dining room table. "Hmm, I need to find something." She walked to the back of the house and went into her bedroom. "Narin," she yelled. "Do you mind throwing something together for dinner with the professor tonight? I need to get some work done."

"Sure," I replied and got to work.

On Tuesday, it was my turn to report for our team at the lab meeting, and we had great data to share. I stepped out of the projector's glare, pressed the button, and said something like this: "Today, I'll be updating you about the work we've been doing, evaluating two putative modulators of DMTA activity on the Type I and Type II receptors. These compounds are H138, which has a novel binding site on the Type I receptor and significantly increases the affinity of DMTA for the receptor, and H140, a newly developed and highly selective Type II receptor antagonist."

I looked at my audience and knew I had their attention.

I continued, "By way of background, on this slide, you see the current model of how DMTA, or the illicit drug Red Sky, is thought to affect the brain. Over-activation of the Type I receptor is considered largely responsible for the euphoria and hallucinations associated with DMTA intoxication. In contrast, over-activation of Type II is linked to mood changes and suicidality induced by the drug. This is called the Cranston-Cohen model."

I reviewed the slides and explained each experiment. The other scientists stayed silent during the presentation. "To summarize, our results show that H138 enhances the effect of DMTA on the Type I receptor, potentially increasing euphoria and hallucinations. The naturally occurring ligand that binds to this site has not yet been identified. However, we believe that developing compounds that block this site could be a useful strategy for treating Red Sky addiction as well as psychotic disorders. Additionally, the compound H140 effectively blocks DMTA's binding to the Type II receptor, which may lower the risk of suicidality in Red Sky addicts. Still, these behavioral effects need to be confirmed."

A scientist named Raman asked, "So, for those who are not well-informed about this field, what exactly is happening with Type I and Type II receptors? As I understand it, these receptors are mostly found in the thalamus?"

"Do you mean biochemically?" I asked.

"No," he replied. "Physiologically, how do they produce these dramatic effects?"

I sat down at the table. "You are correct that these receptors are found almost exclusively in the thalamus, at least in mice and rats. Human studies are still underway. Their normal physiological functions aren't well understood, but they likely modulate the connections between the thalamus and the rest of the brain." I took a drink of water and continued, "The best way to think of it is like this. The thalamus is a relay station heavily connected to the rest of the brain and regulates the typical crosstalk between different areas. For the brain to function normally, these connections need to be well-regulated. Under normal circumstances, DMTA receptors are likely to be involved in controlling these connections. When someone ingests DMTA or Red Sky, it causes over-stimulation of these receptors, likely disrupting these finely tuned connections. As a result, different brain areas struggle to relay information or communicate effectively through the thalamus. This breakdown of interconnectivity allows weak or dormant connections to become active or disinhibited. These connections reach consciousness, where they would normally be suppressed or operate subconsciously. In other words, the brain bypasses the thalamus to perform its functions using ancient connections that humans stopped using long ago. The signals still pass through, but in a new, archaic, and less efficient way. I guess you could call them 'primitive pathways.' The result is that addicts experience a wide range of phenomena. They develop what psychologists call ego dissolution, in which the line between the self and the external world becomes blurred. Sensations like touch, vision, and sound are reorganized and can blend into a phenomenon known as synesthesia. This may all originate from the brain's effort to process sensory input without the help of the thalamus and its network of connections."

I took another sip of water. Morey glanced at his watch. I was well past my time, but I had something important to say. "The result is a feeling that reality, as we usually see it, is falling apart. The Red Sky addict feels a strong connection with things outside themselves. They view the external world as a living being closely tied to their own body. This probably forms the basis for the religious hallucinations that are often described."

I was about to discuss how DMTA could improve cognitive function and ease negative emotions when Morey interrupted. "We're getting off track," he said. "Does anyone have any questions regarding the experimental data Narin presented?"

"I have a question similar to what Raman asked and what Narin is discussing," JoAnn said. "I was wondering if anyone has read the article by Cumberland in the latest issue of the journal *Brain*." Morey grunted from the back of the room, and JoAnn paused briefly but kept going. "He argues that it's misguided to try to separate the symptoms of hallucinations from suicidality. He believes both are caused by the dissolution of the ego, which can result in both intense happiness and rapture but also lead to severe anxiety and paranoia. In the article, if you've read it, he compared it to transporting someone to the top of Mount Everest. It's both exhilarating and terrifying. He cites recent data showing that successful suicides happen even in settings where there was no switch."

Morey yelled from the back of the room: "It's stupid to bring philosophical arguments to scientific questions. Cumberland sits in his office in the Ivory Tower and expels all this bullshit speculation." Morey's voice grew louder and gruffer as he spoke. "If he has a testable hypothesis, then let him propose it. Then we'll test it. We'll put it up for scientific scrutiny using standard methods. If not, let's not waste our valuable meeting time discussing this nonsense. We're scientists here. We deal with facts, not miserable philosophy. We have a theory that the Type II receptor is related to suicidality, and we now have a specific Type II receptor antagonist. Let's push forward with that."

"I think Cumberland might be right," I said. I wasn't sure why I said that. Maybe it was to annoy Morey. Everyone stared at me. "Maybe the suicides have a psychological cause that's unrelated to stimulation of the Type II receptor."

Morey turned bright red. "Are you saying that all your work over the past few years—the Cranston-Cohen model—all of that is wrong?"

I shrugged and sat down next to JoAnn. We had reached an uneasy truce and were working better together. She stayed stiff and formal with me, which was okay, but what I'd said surprised her.

"What was that about?" she whispered.

"I wanted to see how red Morey's nose could get," I said.

We both glanced back at him. He was still fuming, but the redness was subsiding.

In the lunchroom, I saw Carlos and Chris again.

"We were just talking about you," Chris said. "I'm sure you've heard about the recent work on psilocybin."

"The work at Hopkins," I replied unenthusiastically. "Just one dose produces lasting positive effects. It's interesting because it's not the direct effect of the drug, but the actual psychedelic experience that the authors say causes the long-term therapeutic benefit."

Carlos asked, "What do you think about trying that with DMTA?"

"DMTA is too damn addictive." I took a bite of my sandwich. "Once we have the inhibitor of the H138 site, that might be worth pursuing."

Chris glared at me, and I took another bite.

I worked late into the evening and went to the ninth floor to pick up some of my belongings. On top of my desk was a folded piece of paper. I unfolded it and read the note: 'Don't believe anything he says.'

I marched down the hall and slammed on each door with my open fist. "Maru, you little prick, if you have something to say, why don't you say it to my face?" I didn't know which room was his. I pounded harder as I continued down the hall. "You're just jealous of our friendship, you little bastard!"

A silver-haired gentleman opened a door and stepped into the hall.

"Can I help you?"

"Mr. Chandra," I replied, feeling embarrassed upon seeing the company's CEO. "I heard Maru wanted to talk to me, but I didn't know which was his office. I'm very sorry; I didn't realize you would be here so late. I didn't mean to bother you."

"Maru is out of the country," Mr. Chandra said. "If it's important, his secretary sits in that office." He pointed to a glass door at the end of the hall. "She can get you in touch with him tomorrow. You must understand, young man, that one of the burdens of being the boss is that I'm always here at this hour."

"Yes, sir," I replied.

"May I ask your name?"

"Narin—Narin Roy. I work in the Neuroscience Division."

"Yes." Mr. Chandra stepped forward and awkwardly patted me on the back. "Morey has informed me of the excellent work you're doing. Glad to have you on board."

"Thank you, sir," I said. "Thank you."

If Maru's out of the country, then who left the note?

That evening, I finished my book, *Playing Winning Poker*. I had taken detailed notes, and my fingers ached. I wondered when the Desert Prince would visit me again.

I retreated to the basement and spent an hour lifting weights and practicing my jumps. I looked for anything that resembled a knife but only found a dusty hammer in the corner. I held it ready, my legs trembling from exertion, and swung it. The momentum yanked my arm and shoulder as I turned. "I'm not going to let anyone hurt me," I said fiercely. "Just let them try."

Twenty-Eight

On Saturday morning, I visited Sophie's house and gave her a gentle kiss on the cheek. Her eyelids looked swollen, as if she hadn't slept. She stepped aside to let me in.

"Two of my art school buddies crashed in my living room," Sophie said, nodding toward them. "They're driving across the country and staying for a day or two. I told them to stay out of my way. Sarah and Tom." She put her hand on my chest. "How's Dad?"

"He's been in a foul mood," I said. "But I think that's because we've experienced a few delays in completing an important project."

The girl on the couch groaned and stretched. I glanced past Sophie at her. She looked like Sophie but had a rounder face, a flatter nose, and a lighter complexion. When she smiled at me, I noticed multiple piercings in her nose, lips, and ears. Her hair was cut short and dyed jet black. She wore a gray and black striped shirt, a short red skirt, and black fishnet stockings. Tom rolled over. He looked like a pierced, punked-out version of Alex, with a frizzy blond mohawk, gray and black clothing, and a crucifix hanging from a heavy silver chain.

"If you want to learn more about Red Sky addiction," Sophie whispered, "you've got two experts right here."

The four of us had breakfast together. Sophie made blueberry pancakes, and Tom grabbed a beer from the fridge, drinking it straight from the bottle.

Tom said, "Sophie told us you're an expert on Red Sky."

"I study how it affects the brain," I replied. Usually, talking to strangers makes me feel very anxious. But I had changed. Ian had changed me. The Desert Prince had changed me. I was calm. I felt strong and confident. What an incredible change that was.

"Those PSAs they're constantly pushing down our throats are ridiculous," Sarah said. She had a nasal voice. "I don't know anyone who's successfully killed themselves just because of Red Sky. It's like they're totally vilifying the stuff," she added. "I mean, come on. It's no more dangerous than heroin or cocaine."

Sophie said, "I'm pretty sure those are dangerous too. And illegal."

"The point is that all drugs have both good and bad effects," Tom said. "There's no reason to label some drugs as nefarious without seeing any positive benefits. Don't you agree, Narin?"

"I guess so." I watched Sarah as she sliced her pancakes, shifting myself away from the table for a better look at her legs. I couldn't help but imagine how Sophie would look with piercings, a short red skirt, and fishnet stockings.

"So, what were you guys doing in Arizona?" Sophie asked.

"I know it sounds kind of hokey, but there's a guy who calls himself Baba," Tom said. "He's got this ashram in Sedona. It's gorgeous. He always has people from all over the world and lets them stay there for free, feeding and teaching them. If you guys are interested, I'd definitely recommend it." He took a swig of his beer and continued, "Baba teaches this philosophy called The Coming Ethic. Basically, he says ethics shouldn't be about how men treat each other but about how men treat Mother Earth. We're all products of Mother Earth, yet we act as if she belongs to us, like she's our property or something, not the other way around. We're headed for a terrible environmental catastrophe where the economy's going to collapse, and almost everyone's going to die. Mother Earth is reasserting her authority over us. Unless we learn to live in harmony with each other and the other creatures inhabiting this planet, our species is doomed." Tom took another drink. "That's what he said, and I believe that too."

"They had the most beautiful art, Sophie," Sarah said. "The walls were covered with gorgeous landscape paintings of the red rock canyons. You would've loved it." Sarah looked at Sophie. "You're really limping," she said. "Does it hurt to walk?"

Sophie sat down and tried to smile. "I'm fine," she said.

Tom and I played cards while Sophie showed her paintings to Sarah.

"What's so funny, dude?" Tom asked.

"I was just thinking about how lucky I was the other night."

"That's awesome," Tom laughed. "With Sophie or someone else?"

"No, in poker," I explained. "I played this shitty hand. I could have lost my shirt. But somehow, I ended up winning." I was eager to share my story, and since I knew Sophie wouldn't be interested, I told Tom. "I was holding a pair of fives in the pocket, and then the flop had two cards higher than mine. I don't know if you know much about poker, but that's a pretty bad hold situation. But everyone else stayed in, and I had three cards to a straight and three to a flush, so I decided to stay even though the odds were against me. Anyway, the turn came up a five, and the river was a five. So, I ended up with four of a kind. And since everyone had stayed in, I made a bundle, somewhere around three hundred bucks.

"That's so wicked. High-five, man."

"Excuse me," Sarah announced. "Sophie claims that your loud chatter is distracting her. Everyone needs to be quiet. Ms. Picasso is trying to paint."

"Speaking of Picasso, where the hell is Pablo?" Tom asked. "I've only seen that other fat black cat."

"She's hiding somewhere."

Sophie stood in front of her blank canvas, her face tense with concentration as she made the first stroke.

"It's so funny to watch her paint," I laughed. "She's like a warrior in a mortal struggle with the canvas."

Tom snorted and knocked over his beer.

I continued, "When she carves, she dances, whistles, hums, or sings along with the radio. But no, not when she's painting; she's drenched in sweat and covered in paint. She's got streaks across her face. She looks like an Amazonian princess."

"Well, as Ravi used to say, 'Of all the arts, painting is the one that most requires us to create something out of nothing. That's what makes it so challenging.'"

"Oh, so you knew Ravi Reddy, too?" I asked.

"Everyone had to take his class. It was required," Tom explained.

I lowered my voice, even though Sophie was too far away and too focused to hear or care about what I said. "You know about what happened with Sophie, right? About the accident?"

"I'd heard about it, but it didn't quite make sense," he said. "Sophie hated Ravi."

"Well, sometimes there's a thin line between love and hate," I said.

"The other strange thing is that Ravi was gay," Tom said. "The last time I saw him, about three years ago, he was all over some high-powered lawyer at a drug company. I know everyone was high at the party, but their behavior was totally inappropriate."

"A lawyer?" I asked. "What did he look like?"

"He was a skinny blond guy with blue eyes. He had on a really expensive suit, and his blonde hair was slicked back. Ravi introduced him as his lawyer and his muse. That's how I knew he was a lawyer. I never found out his name. But from the way Ravi talked, they seemed really into each other. I'm not gay or anything, but I could see why Ravi was into him."

My heart was pounding wildly in my chest. "Does Sophie know that?"

He shrugged. We both looked over at Sophie, whose face was twisted in concentration. "I don't know when she started her fling with Ravi," Tom said. "But here's the weird thing that I can't get my head around. Sophie hated Ravi for good reasons. He was always picking on her, I mean, really tearing into her, giving scathing reviews of her work right in front of the class. He was rude to others, but he outright mocked her and told her she was worthless. Why would she want to marry an asshole like Ravi? I guess he must have made her an offer she couldn't refuse."

When Sophie finished painting, I walked over to her and massaged her shoulders as she cleaned the paintbrushes.

"I'm finished with my daily penance," she said. "But I can tell from your expression that you're not. What's wrong? Is Tom driving you crazy?"

"Actually, I'm learning a lot from Tom."

"Anything I might be interested in?"

"I'm not sure." I kissed Sophie on the cheek; she smelled of paint and sweat.

Tom rolled a joint and sat on the porch, smoking. I sat beside him.

"I'd forgotten about the pea soup summers here," Tom said. "I don't know how anyone can live here. You only get about five decent days a year. The rest of the time, it just sucks." He slapped a mosquito on his arm. "You want a hit?" Tom offered me the reefer.

I shook my head. He took a long drag.

"I was just contemplating what an incredible time we live in," Tom mused. "With a pill, you can reach the highest level of ecstasy. You've got uppers to keep you awake and downers to bring you down. You've got the internet and virtual reality. Soon, there'll be nothing we won't be able to experience. There's no need for religion or philosophy anymore." He puffed on his joint and continued, "But most of all, there'll be no need for art in the traditional sense: paintings, sculptures, plays, books, music—you name it. You pharmaceutical assholes have killed off the need for any of the shit artists create. People like Sarah, me, and even Sophie will become completely obsolete. We're like dinosaurs headed for extinction. We might as well all just put a gun to our heads and get it over with right now."

"Are you serious?"

Tom took a long drag and closed his eyes. "Serious as all fuck, man. Serious as all fuck."

"Here, give me a hit," I said.

"It's laced with Red Sky," Tom warned. "I micro-dose the stuff with my reefers so I don't have to take all those psychiatric medications and be a zombie all the time."

I signaled to Tom to pass me the joint. Once I had it, I brought it to my lips and took a hit. My lungs constricted with pain. I remembered an adage from pharmacy school: 'Everything is a poison. It just depends on the dose.'

I spent the night with Sophie. She wrapped an arm around me as she settled in. My mind was a whirlwind, trying to process everything Tom told me. The 'lawyer dude' was definitely Ian. It probably started as a professional relationship, with him handling legal work for Ravi, but it quickly became much more serious. Ian told me he joined Harvester five years ago, so the timeline fits pretty well. Sophie said they had a car accident two years ago. So, something must have changed between that drug party where the two boys were hanging onto each other and the car accident.

"Was Ravi a good teacher?" I asked. I wasn't sure if she was still awake, but she was.

"He was pretty demanding," she said. Her apparent indifference to his abusive behavior made me angry.

"Tom said he really abused you in his class," I said, trying to hold back the bile that was building inside me.

"He had his days," she said, holding me tighter.

What the hell, Sophie? I couldn't understand how she could be so dismissive of his abusive behavior.

"He thought I wasn't working up to my full potential," she said. "That's what he told me when I got to know him better. That's why he was always on my case."

"He told you that?" I asked. "That's why he said you would never make it as an artist?"

"I wouldn't have," she replied. Her calmness was driving me crazy. "If he hadn't lit a fire in me, I wouldn't be able to paint like I do now."

I wanted to throw her arm off me and storm out, but I didn't. I just lay there until I fell asleep. I had some sexy, wild, crazy dreams that night, and they all involved Sarah.

The next day, Tom and Sarah packed up and left. With a quick hug, I felt Sarah's cool, soft skin and inhaled the scent of her hair — the very things I'd spent the night dreaming about.

"They seem nice," I told Sophie.

"I'm glad they're gone. Now I can finally get some work done."

Sophie worked the rest of the day while I prepared the meal to take to the professor's house. Since my recipe options were limited, I looked online for something simple. I chose portobello mushroom quiche because it seemed easy, and we had all the ingredients on hand.

That evening, we went to Goldblum's house with a quiche in hand.

Sophie said, "Sarah offered me some Red Sky last night." I could tell from her tone that she was joking.

"Did you give it a try?" I asked.

"No, I just freebased some crack cocaine instead," she said, deadpan.

"Good choice," I said.

"I'm not saying Red Sky is more dangerous than crack," she said with a laugh.

"Of course, it isn't. That would be totally unfair."

Twenty-Nine

As June ended, the nights no longer cooled to a comfortable temperature. The professor never turned on his AC; instead, he used a slowly rotating fan that circulated the hot, stale air. Oddly, he seemed unaware of the heat. He always wore the same clothes: khaki pants, a dress shirt, and a sweater vest. I told Sophie that the professor's glasses were filthy. She discreetly picked them up and wiped them clean with a dry cloth.

During dinner, I couldn't take my eyes off Sophie. Maybe Sarah played a part, stirring my hormones. I kept admiring the lines and curves of Sophie's slim figure and her radiant skin. Her words washed over me. The lilt and melody of her voice, along with her beautiful face, held my attention. I couldn't wait to get her into bed.

After dinner, we headed back to her place.

"Weren't you hot in there?" I asked.

"He gets cold so easily," she replied. "He hasn't had the energy to go for his morning walks lately, so I insisted he see a doctor. He said everything checked out. Last year, he had terrible pneumonia and ended up in the hospital. I'm worried about what will happen to him this winter."

"Does he have any family?"

She pursed her lips. "He never mentioned any to me."

"Then we need to make sure he writes his will before winter. If he leaves the house to me, I can live across from you."

She glared at me. "That's not funny."

I should have known better. She loved the old professor.

We went into her house, and I put away the dishes.

"I just hope that when I reach his age, I'm as independent as he is," she said. Then she read me a long passage from the professor's book about how those who want to create must first decide what they want to destroy, or something like that. I wasn't paying attention to what she was saying; honestly, my mind was on sex, and nothing else really mattered.

"If only we could do that," she said wistfully. I moved behind her and grabbed her. I thought about taking her to the bedroom and having my way with her. She set the book down. "I'm sorry, Narin, but I've got to get back on schedule. I'm so far behind now." Her voice broke.

"What's wrong?"

"I'm just really stressed," she moaned. "I suck at painting. I don't know who I'm trying to fool."

"What're you talking about?" I craned my head to look at her beautiful face. "You're fantastic. And don't worry, I'll help you in any way I can. I'd do anything for you, you know that. I've been thinking about giving you some money. I've got plenty now, working for Harvester."

That was the wrong thing to say. She gave me a fierce, almost hateful look.

I stepped back as if I were avoiding a deadly blow. "If that would make you happy," I said hesitantly.

She kept glaring at me and wiped away a solitary tear. I wasn't sure what had triggered her reaction. After thinking about it, I realized my mistake. She didn't want to rely on me or anyone else for her success. She had leaned on Ravi, and that hadn't worked out. Now, she was determined to succeed on her own. She was completely unaware that I could be her greatest asset. I decided to let it go; there was no point in trying to reason with her. She stomped off and started working. I sank onto the couch, lost in my thoughts.

I began reflecting on the lonely times I spent in my room as a kid, wishing I had friends to talk to. But along with those memories, I could hear the beautiful sounds of Bengali in my mind—conversations with my mom, dad, uncles, and aunties. I hadn't realized how much I missed hearing Bengali until that moment.

I thought about my dad and how I learned to play tennis with him when I was a kid. He would gently lob the ball to me until I finally figured out how to hit it back. After a while, I could play reasonably well and even enjoyed our games together. But we stopped playing when I was a teenager. Who decided to stop? I couldn't remember. Maybe it was a mutual decision. Or perhaps it wasn't a decision at all. I remembered how my sister and I played together as kids and how I pretended my belt was a giant snake that held her dolls captive in its secret lair, only for Superman and Luke Skywalker to defeat the wicked monster and rescue the captive damsels.

Sophie finished her work around midnight, took a shower, and brought over a half-empty bottle of wine. Her wet hair hung over the back of her thin T-shirt. She lit a candle and settled onto the couch next to me.

"Feeling better?" I asked cautiously.

"I'm exhausted," she declared. "Here, massage my arms."

I followed her instructions and couldn't help but marvel at the contrast between my dark skin and her pale complexion. What a strange and beautiful difference. I guess she prefers dark, handsome guys. That made me think of Sanquel. He was just as dark as I was and much more attractive. A wave of jealousy washed over me again, but I made sure not to show it to Sophie. I had been feeling a lot of hostility toward her and didn't want it to ruin what was about to happen.

We talked, and soon we were whispering to each other. Her skin felt cool against my warm hands. The burning candle filled the room with the scent of cinnamon, which surprisingly did not bother me at all. I told her how, as a child, I'd pushed a pencil tip through the holes in my bedroom window screen. The holes grew large enough that one day a colony of ants entered through them and built a home under the table where my terrarium sat. My mom was furious.

The rain began to fall. The curtains swayed, the flame flickered, thunder rumbled, and raindrops pounded against the roof.

She took my hand and led me into the bedroom. I was on top of the world.

Afterward, she curled up in bed, and I held her close.

"Are you in pain?" I asked.

"A little." She arched her back.

I pressed my hand against the small of her back and gently rubbed. I had tried to be as gentle as I could.

"Ian's so funny," I said. I don't understand why I kept thinking about Ian, but I couldn't help it. He was never really far from my mind. And after what Tom told me, my obsession with him only worsened.

She remained silent.

"He says that if I stay at Harvester, he'll look after me. In a few years, I'll be VP of Drug Development, and we'll have enough money to do whatever we want."

She stiffened at my words but remained still. My hand rested on her back. "It'll be good. We'll be happy. Our kids won't have a care in the world." Shit, I had no idea why I was talking about kids, but I was so in love with her that I couldn't imagine a future without her. Still, I'd never told her I loved her. I was close to saying it, but the words wouldn't come out. It felt like crossing a line to a place from which you can never return. But I was ready now.

Before I could say anything, she said, "With my injuries, I wouldn't count on it, kids, that is."

Her breathing grew heavier as she curled into a ball, shoulders shaking while she tried to hold back her sobs. I didn't know what to say to comfort her. I guess I could have told her that it's okay; I didn't really want to have kids. I felt sorry for her. Did I really want kids? I hadn't thought about it much before. I assumed I'd have kids someday. I needed to think this through later.

I gently stroked her shoulder while my mind stayed on Ian.

Ian and Ravi were more than just friends. Ian was probably in love with him. When Ravi died, Ian must have been devastated. Maybe he felt guilty somehow. Could he have been responsible? I heard that Ravi's art in Harvester's gallery belonged to Ian, not Harvester, and it was now worth a small fortune since Ravi had died. Perhaps his love for Ravi had shifted to Sophie. I didn't know what Ian's sexual preferences were, but I could tell from the way he looked at her the other day that he found her attractive. But if that's the case, why would Ian involve me?

'Don't believe anything he says.' Those cryptic words kept echoing in my mind. What could Ian's endgame possibly be?

I could tell Sophie was asleep from her breathing.

I was utterly dependent on Ian. But what was motivating him? I needed to figure out Ian's game. Then I'd beat him at it. It's like poker; it's very psychological. I would outplay Ian and come out on top.

I couldn't sleep, so I pushed the covers aside and stepped out into the warm night. The sky was overcast and silent, with no stars. It was a good night for thinking. But my mind was just as blank as the sky. I splashed through dark puddles, feeling like a tiny gear in a vast machine set in motion long ago. I was powerless to stop the relentless push forward. Where will this all lead?

I turned back toward the house, feeling disappointed and frustrated, when two arms reached out from the darkness and grabbed me tightly around my shoulders and neck. They clung to my collar and dragged me down. All my training from the Desert Prince abandoned me. I struggled and gasped for air. The arms were strong. A broken voice spoke desperately: "Have you seen the light? It is all around us. Open your eyes and see!" The grip loosened and released me. I coughed and sputtered, pushing myself off the ground and brushing myself off. A man lay on the ground, staring vacantly and muttering to the night.

"Fucking Red Skyer." I wanted to kick him—kick him in the stomach, in the face, in the groin—until he screamed like bloody death. But I didn't.

I returned to Sophie and slid back into bed beside her. Her breathing was slow and steady. I gradually relaxed and drifted off to sleep. The last words I heard as I fell asleep were, "Open your eyes and see!"

Thirty

The next day, I received a text from Deepa. Our parents planned to return from India in July. They invited Rajeev for dinner and wanted me to come home so I could be there. She didn't ask how I was doing or anything else. I guess she was still upset with me.

As our project neared completion, Morey moved to the third floor and settled into my office. There, he examined every detail of our work, reviewed the data, and muttered to himself. He was brooding, intimidating the technicians with his commanding orders, snapping at the slightest delay, and mostly pacing back and forth while running his hands through his unruly hair.

"He needs everything to work perfectly," Ian said, finished his coffee, and set down his cup. "He's bet everything on this project. There's a big payoff, but only for unequivocal success."

"I don't see why it wouldn't work," I replied. "The data looks good."

"We'll find out soon after we deliver our product to them. They're eager to start using it. Our contacts say they'll report back in two weeks after the first tests to let us know if our concoction worked or not."

I stared at him. "We'll know that soon?"

Ian went over to his bird. "I've been trying to teach her to say 'Nevermore,'" Ian said and held the mealworm high. "Say 'Nevermore,'" he said in a high-pitched voice. The bird let out a deep croak and stared at the wriggling morsel.

I stirred my coffee. *What's your game, Ian?* I nearly said it out loud. I wanted to know what he was hiding from me and the truth about him and Ravi. Something in Ian's manner, tone, and word choices had always made me suspicious, but the feeling was so faint that I ignored it. Still, as I watched him, I realized I had been lulled into complacency.

It almost felt like I was seeing Ian for the first time as I watched him feed his bird. I realized that everything he said and did was meant to create an effect. The signs were clear: how he rolled up his shirt sleeves, leaned casually against the kitchen doorframe while talking to Sophie, or that amused smile just before he burst into a hearty laugh. It was all rehearsed and meant for an audience. I was his audience. It all seemed fake, yet he made it appear very real. But now I was onto his game.

Ian interrupted my thoughts. "I got a call yesterday from a friend of mine from law school," he said. "He's working for a small startup in San Francisco and wants me to join him out there."

"Is that so?" I took a sip of my coffee.

"They've developed an implantable device to help people who always feel tired; 'excessive daytime somnolence' is what he called the condition." Ian dropped a mealworm into the bird's mouth. "If it works, it'll be in high demand. It mimics the sensation of a pinch by delivering random shocks to a nerve in the forearm."

"Does it hurt?"

"He says people don't perceive the shocks as painful, just a slight buzzing. But the body reacts as if it were a sharp pain. It sends a jolt to the brain's attention centers and wakes you up. Does that make scientific sense to you?"

"I guess so," I replied. "I imagine people would get used to it, so it wouldn't last very long. People can get used to anything."

"He says their data shows it's very effective and doesn't have side effects," Ian said. "They're working on getting expedited approval from the FDA. That's what he wants my help with."

"So, are you leaving Harvester?" I asked.

Ian laughed. “I told him that I couldn’t do that right now.” He came and sat down beside me. “They treat me like an old car around here.” His face tightened. “They don’t care what kind of damage they do to me. They’re just waiting for me to break down and leave.” Ian clenched his fists. “They don’t understand that even if I’m not worth much, replacing me will cost them a hell of a lot more.”

More pretending. You should have been an actor, Ian. You almost had me convinced.

Ian glanced at the clock. “You should get back to work. Morey will wonder where you are.”

I finished the day late on the ninth floor. At Morey’s insistence, I reviewed all the data from that week’s experiments to look for errors. As I wrapped up, I heard voices from Ian’s office and went over to listen. I grabbed my tumbler and pressed it against the wall.

Someone said, “We have a problem.”

I wasn’t sure who was speaking, but I heard Ian call him “dad.” I remembered that Ian’s father, Harold Blair, and Sanjay Chandra were college friends, and Harold served on Harvester’s Board of Directors.

Harold said, “Chandra wants to shut it down.”

“Now?” Ian asked incredulously. “At the last minute, he’s getting cold feet?”

Harold said, “He was never really on board. His conscience is catching up with him.”

“What the hell?” Ian said.

“You have to understand, Ian. He saw things during that first trip to South America that left him psychologically scarred,” Harold said. “He and Morey received a warm reception from the state governor back then. For a substantial bribe, they got assurances they wouldn’t have trouble collecting medicinal plants from the indigenous tribes. But he soon realized what that really meant: people were beaten and hanged for not cooperating. It really affected him.”

“But that’s exactly why we need Operation Juggernaut,” Ian said. “So, you don’t have to resort to beatings and torture.”

"I get it," Harold replied. "If this works, there isn't an army or police force anywhere in the world that wouldn't pay top dollar to get their hands on this stuff. Just imagine how easy it will be to get confessions from criminals."

"I don't understand his gripe, then," Ian said.

"He wants it shut down," Harold replied flatly. "He wants the contract canceled."

"And he's willing to sacrifice Harvester because of his moral qualms?" Ian shouted. "Because, you know, Harvester will be swallowed whole by a big multinational corporation if the stock price drops any lower."

Harold lowered his voice, but I could still hear him. "We just need to keep up the pressure a little longer. Make sure that Alvarez gets whatever he needs to keep the lawsuit going." I had never heard of Alvarez before, but it only took me a minute on Google to find out he was the lead attorney representing the indigenous tribes in the lawsuit against Harvester. It was the lawsuit that had caused the fall in Harvester's stock price. Harold and Ian were betraying the company.

Harold pressed on, "With Harvester's stock price in the gutter, the Board is done with Chandra. He'll be gone soon. Once he's out, we can finalize things with Alvarez. Then everything will be fine, and we will be in control."

"Sanjay was fine in the 1990s," Ian said. "But now, this place is a wreck, with each division operating in silos. And, God, can you imagine what would happen if Maru took over!"

"I'll handle Sanjay," Harold said. "He knows his time is up. He'll quietly retire."

Ian didn't reply.

"What about Maru?" Harold asked. "He could cause trouble for us."

"I've got so much on him that I could have him locked up for life," Ian said.

"Good," Harold said. "So, who is this Narin anyway?" He didn't pronounce my name correctly, but I knew who he was talking about.

Ian said, "He brought us the Type II receptor blocker."

"Yeah, but does he know about Operation Juggernaut?"

Ian didn't reply immediately. I prayed he'd say, 'No, Narin is just some low-level idiot.'

"He knows," Ian said. "I told him about it. He's with us."

"What the fuck, Ian?" Harold yelled.

I agreed. What the fuck, Ian? Was he trying to get me killed?

"I trust him," Ian said.

"You know, we had to go through a hell of a lot of security clearances just to find out about Operation Juggernaut," Harold said. "You can't just go around telling your friends about it."

"You're right," Ian said. "But I trust him. I guarantee he won't cause us any trouble."

"What about Morey?" Harold asked. "He's a loose cannon. He always has been, and he always will be."

"I'll take care of Morey," Ian said. "So, will you be the interim CEO when Chandra is forced out?"

"That's the plan," Harold said.

They stayed silent for a moment. Finally, Harold said, "You don't show up at Cardinal's games anymore, Ian. Harvester has a suite at Busch for a reason."

"Is there a game on Saturday?" Ian asked.

"There is, and I'll be there. If you go, I'll make sure you have all-you-can-eat nachos available. You like nachos."

"I liked them when I was ten, Dad."

They both laughed.

"Come here," Harold said softly. I imagined them hugging.

"We'll get through this," Harold whispered. "We're almost there."

After Ian's dad left, I sank to the floor, still holding the tumbler.

What the hell have I gotten myself into?

Part IV

Thirty-one

It's already morning. The light in the cell hasn't been turned on yet, but the room is starting to brighten. There are no shades or bars on the window. I haven't given much thought to exactly where I am. The windows are too high to see out and figure out my surroundings. The incident happened at Harvester, so I might be at the Chesterfield Police Department. Or maybe they took me downtown because of the nature of the crime. I'll need to ask the next time I get a chance.

I've noticed that the effects of the Red Sky are starting to fade, and my mind is becoming fuzzy. Fortunately, this section covers recent events, so I should be able to provide an accurate account.

I'm nervous about meeting the FBI and DEA officials today. I'll be honest with them, but I won't mention anything about Operation Juggernaut. If they ask, I'll pretend I don't know anything. The less I reveal, the better off I'll be.

When I joined Harvester, I never imagined I'd get caught up in this life-and-death drama. First, there's my role in Operation Juggernaut, which, after some mental gymnastics, I started to accept. Then, I discovered Ian and Harold Blair were planning to take over Harvester! Not only that, but they also orchestrated the company's decline to push Sanjay Chandra out. They'd swoop in, take control, and 'clean up the mess.' If their plan succeeded, Operation Juggernaut would make Harvester—and themselves—a fortune.

I thought through the possible scenarios:

Scenario 1: If Ian and Harold fail, Ian will lose his job. Then, all of Ian's promises to me would be pointless. My future would rely on Morey, who lied to me at our first meeting. I'd have to trust his word to get credit for finding the Type II receptor blocker. But what about Sophie? If Morey found out I was sleeping with his daughter, he'd fire me right away. I can see him, red-faced, telling me to 'get the bloody hell out of Harvester and never come back.' I'd end up with nothing.

Scenario 2: Harold and Ian successfully carry out their coup d'état and take control of Harvester. The Chandras sail off into the sunset. Operation Juggernaut is a huge success, rejuvenating the company and providing the resources for it to become a major player in the pharmaceutical industry. We develop a blockbuster drug to solve all the problems related to taking DMTA. Ian makes sure I get credit for my work. I climb the corporate ladder and become a renowned researcher. I support Sophie and help fund her rise to stardom in the art world.

Those were the only two outcomes I could see.

I guess I could go public with what I knew about Operation Juggernaut, but that got me nothing and put me at risk of being killed by whoever was behind it.

No, I had to stay fully committed to Team Blair, and I genuinely hoped they knew what they were doing.

Let's return to the story before I forget it all.

Ian invited me to dinner on Wednesday evening, probably during the last week of June. I was still stunned from overhearing Ian and his father talking. I had a lot of questions, but I couldn't reveal that I'd been spying on them.

I locked my laptop and lab book at the end of the day and was about to leave when I saw Morey sitting alone in the corner, clicking through graphs on his computer screen. Something about his posture made it seem as if he were waiting for me. I walked over, but he kept looking at the graphs.

"It looks like you're on track," Morey said. He pointed to the graph on the computer screen. "You need another control here. A couple of your experiments are missing proper controls."

"That data is still pending," I said. "We'll have it by the end of the week."

Morey didn't reply.

"I'm sorry, Morey," I added. I hated disappointing him. He had become a father figure, and I craved his respect. I'd come to trust his judgment on all scientific matters. I began to think he was the most brilliant person I'd ever met. I just wished he would be okay with me and Sophie being together. But, at that moment, that was too much to ask.

"We'll stay on track from now on," I said.

Morey folded his hands in front of him. "There's a bloody lie that we're told from childhood," he said. "It's fed to us from everywhere and everyone. If you work your ass off, then you'll end up rich, happy, and successful." He turned to me and continued, "But that's not the case, Narin. It's not the workers who benefit from their hard labor." He shook his head. "Did you know that bastard Ian just got approval for another assistant and two more attorneys in his little fiefdom?" Morey shook with rage. "I don't understand why Sanjay puts up with that little prick. All Ian does is strut about like a gamecock while his underlings do all his work. And then he swoops in at the last minute and takes all the credit."

Morey paused for a moment, lost in thought. "But I don't care about all that. I don't need an assistant. When have I ever needed or asked for one? I can stay here all night, getting my work done. Everyone knows nothing is waiting for me at home." Morey went back to staring at the screen. "But it would be nice to get an occasional word of acknowledgment—at least a thank you, a pat on the back—for helping build this company brick by goddamn brick over the last twenty years. That certainly would be nice."

"Goodnight, Morey," I whispered. Morey kept staring at the screen. I wished I could say something helpful, something to comfort him, but I had nothing.

I met Ian in his office, and we headed to the lobby together.

"Good night, Sue," Ian said to the receptionist, who was working late that evening.

"Nighty-night, Ian," she said, looking up and noticing me. "I mean, Mr. Blair." Her voice quivered. "Have a good night, Mr. Blair."

Ian chuckled as we left the building.

"She's quite pretty, don't you think?" Ian asked. "She's popular with the residents of the ninth floor."

I drove and briefed Ian on our DMTA work. Ian nodded but didn't ask for any details. He pointed and guided us as we headed downtown.

"Where are we going?"

"It's a place called Citadel. I know this place, and they know me," Ian said. "It's got an eclectic selection. I'm sure you'll find something you'll like."

I had so many questions for Ian, but I couldn't figure out how to ask them without revealing that I'd been spying on him and his dad. Not to mention all the things Tom had told me about Ravi and Ian being in a relationship. *How many secrets can one man keep?* Plus, I needed some guarantees from him. I had to find a way to get these promises out of him. I considered different ways to bring all this up at dinner without completely pissing him off.

Upon arriving, the greeter and server approached Ian with a mix of familiarity and formality. We bypassed the waiting crowd and were led to a gray metallic table, faintly illuminated by a red lantern. A pretty girl with short brunette hair streaked with purple handed each of us a menu. She wore a cropped black shirt. I noticed the shiny silver star piercing her navel. "Hey Ian, how have you been?" She leaned forward, causing her shirt to fall open, revealing a black and red bra. She wrapped her arm around him.

"This is my friend Narin," Ian said, nodding toward me.

She looked at me, her body still pressed against him. "It's nice to meet you. I'm Jill. I'll be your server today." She held Ian and whispered in his ear, "So, how's life in the fast lane?"

"Very fast," he replied with a smile. He picked up the menu and pointed to something on the wine list.

"You got it," she said brightly, releasing him. "I'll be back to get your orders."

Ian watched her strut away. "She must be available again. That's good to know."

Ian's eyes locked on me. "I can tell you've been working out. You're looking great." He flashed a smile. "So, how's everything with Sophie?"

"Good," I replied.

"Have you taken the poor girl out on a date?"

I ran my fingers through my hair. "Not exactly."

Ian frowned at this. "Must I teach you everything? You need to take her out and show her a good time. Go somewhere where the two of you have to dress up."

"She enjoys being at home, and I have to admit, I don't mind it. She has a very comfortable couch."

Ian leaned forward. "You might think you don't need any glitter or extravagance, but you do. Don't fall into the rut of an old married couple. There'll be time for that later."

"I don't know," I said. "It's been so hot lately."

"Listen," Ian said sharply. "I've got season tickets to the Fox. I'll give you two tickets for this weekend that I won't be able to use. I'm not sure what's playing, but the shows there are always good. Take her out. Show her a good time."

I didn't like where this was heading. "Maybe you should take her out," I said. "You'd probably be better company than I am."

Ian looked at me, his bright blue eyes studying me. "You've taken my advice," he said finally. "You haven't taken her on a date, but you've made progress with her. It's clear as day."

I glared at him and tugged at my hair. "What makes you say that?"

"I can tell," Ian said. "You're not good at keeping secrets, and I'm happy for you."

These words painted a mental picture of Ian peering through the blinds while Sophie and I were having sex. Even though I doubted he really stood outside Sophie's bedroom window, he somehow knew about her movements, and now mine.

Ian picked up the menu. "Everything here is great. I like steak myself. Choose what you like. Of course, dinner is on me."

I scanned the menu but couldn't focus. My eyes landed on the restaurant's name, 'Citadel.' What is a citadel? Is it a castle or a fortress? The word made me think of a protective structure, the part of the city most resistant to attack. I pressed my lips together and looked at Ian again. "Since you seem to know so much, Ian, maybe you can tell me if things will work out between me and Sophie." I was curious about what he'd say to that.

"Why?" Ian asked. "Do you foresee problems?"

It was time to be honest. I was completely in love with Sophie, but I still had lingering doubts about a long-term relationship with her. My mind was a jumble of uncertainties. What would my parents think? Shouldn't I look for someone more educated? If I have kids, I want them to be smart and successful. Hell, I wasn't even sure if Sophie could have kids. I should probably find an Indian girl, someone my parents would approve of. Maybe I should let my parents find someone for me and just get it over with.

My eyes kept scanning the menu. "We're just different, that's all, and I barely know her," I said. "All I'm saying is I'm just not sure things will work out."

Ian nodded thoughtfully. "I certainly hope things do work out between the two of you. I don't think you realize how sincerely I wish for you and Sophie to be happy."

"You've said that before," I said. I felt especially confessional that evening. "It's strange, Ian. Sometimes, I think she's two different people."

"Two different people?" He looked confused.

"When she's working—and she's almost always working—she's all business, serious, almost fanatical. But when she's had too much wine, she gets this strange, mischievous look in her eyes and becomes playful, carefree, maybe even careless. I don't know if I can handle one of her, let alone two."

I felt anxious talking about all this, so I fiddled with my phone to keep from pulling my hair out. "But I'm glad you're on my side and that you seem to care so much," I said. "It's just that she's not the kind of girl I thought I'd end up with."

A twisted half-smile appeared on Ian's face. "What kind of girl did you think you would end up with?"

I shrugged. "I don't know, maybe an Indian girl, a doctor, or a scientist."

Ian kept studying me: my drooping shoulders, tired eyes, and pressed lips that constantly formed a frown like my mother's. He opened his mouth to speak but then changed his mind.

Jill came back, set a bottle of wine on the table, flipped the glasses, and poured drinks for both of us. We thanked her.

"Let's have a drink," Ian said, raising his glass. "Frankly, I don't trust people who don't drink. People who refuse to drink have secrets to hide, and there shouldn't be secrets between good friends like us. Cheers." *No secrets, indeed.* We both took a drink.

Jill came back, recited the specials, and we placed our orders. Ian ordered the filet, and I went with the sea bass.

"Jill reminds me of the first girl I ever kissed," Ian mused. "Her hair smells like grapes, and her clothes smell of strawberries." Ian licked his lips. "If you don't mind me asking, what does Sophie smell like? I've always been curious about that."

I couldn't help but laugh. "You're a bit of a voyeur, aren't you?"

"I'm interested in people," Ian said. "But that can get you into trouble these days, being too interested, especially when you're in a position of power. They call you a creep, a pervert, a predator, or even worse. But you have to admit that there's a certain thrill in seeing the intimate details of people's lives." He took a drink. "And seeing all their pain. Maybe that's just me. Like I said, I'm interested in people."

I didn't say anything; I just kept fiddling with my phone.

Ian continued, "I've started a practice of holding the gaze of every girl I meet, not turning away until she does or if she gives me a gesture of acknowledgment, like a smile. It can be quite uncomfortable if you're not used to it. I try to focus on something, like determining the color of her eyes."

"How does that work?" I asked, genuinely curious.

"I get a few smiles," Ian said. "Many girls will look away in disgust as if you've insulted them just by looking in their direction."

"You must run into a lot of Indian girls then," I said somberly.

Ian chuckled and took a drink. "So, let's hear about Sophie," Ian said. "You didn't answer my question. By the way, I tried that with Sophie, staring into her eyes. She held my gaze for an admirably long time before turning away."

"I noticed you were staring at her," I said. I wanted Ian to know I'd seen it. "To answer your question, she smells clean, like a girl. That is, except when she's been working. Then she smells like sweat and paint thinner."

"You can tell a lot about a person by how they smell," Ian said. "At least, that's been my experience."

"Well, I hope I smell okay."

Ian laughed. "You have the smell of success on you, Narin. That's one of the first things I noticed about you."

The entrées arrived.

I finished my first glass of wine faster than I planned. I guess I was pretty thirsty. As I started sipping my second glass, I swayed in my chair, feeling a bit buzzed. I thought about Sophie and how she drinks alcohol like soda—quickly and in big gulps.

"Thank you for inviting me to dinner today," I said. "When you asked me to dinner, I thought you might invite Sophie too, since you seem fond of her."

Ian nodded. "I considered it but decided it would be best not to do so. She's your friend, not mine, and I believe she doesn't care much for me."

"I think she gets nervous around people she doesn't know. We went out with my housemate and his partner the other day, and she hardly said a word."

"Well, maybe if she gets to know me better, she'll see my good side," Ian said with a smile. "Regardless, I'm glad to have you here with me today. Nothing is more important than friends in this world. Experience has shown me that people often treat friends with more kindness than their own family." Ian cut his steak. "You need to look no further than your friend Sophie to see proof of that."

I pointed a fork at Ian. "Now, that's not fair. I know exactly what you mean. I'd be the first to admit that she should reconcile with Morey. But if she did, that would be the end of our relationship. So, if you care about us, as you say you do, then you'd know they should stay estranged."

"Why is that?"

I struggled to enunciate my words: "Because, as you know, Morey hated Ravi. And apparently, Ravi was my twin brother."

"He was your twin brother?" Ian asked with a mix of amusement and confusion. "Now that you mention it, you do resemble him a little. I hadn't noticed that before."

You might be surprised that I've never looked up a picture of Ravi Reddy. I just held onto this mental image of a different version of myself—one who is more successful, self-confident, and older.

Ian pressed on. "Are you saying it was a racial thing? Morey doesn't like Indians?"

"Morey hates Indian people," I stammered, making a sour face. "He detests them." I was buzzed at that point and didn't know what I was saying. I'm a lightweight when it comes to alcohol. It doesn't take much to get me drunk.

"Maybe he had other reasons," Ian suggested. "Ravi was a fifty-year-old man screwing his twenty-something daughter."

"It wasn't just that, Ian. Morey didn't like his brown ass," I growled loudly enough that the couple at the next table looked over. "He would never let me marry Sophie. He'd do anything to stop it. I know that for a fact. She told me so. Morey would kill me before he let me marry his precious daughter."

Ian shrugged. "Well, if that's the case, then we have no choice but to kill Morey first."

I stared at Ian. *What was he saying?* "We'd have to kill Morey?" I whispered.

"I'll hold him down, and you shoot him in the head," Ian said. The steak oozed red juice as he sliced it.

"I couldn't do that, Ian."

Ian glanced at me. "You couldn't? Not even for love?"

"It's not my style." I picked up my knife. "I'd stab him in the heart instead." I laughed uncontrollably and poured myself another glass of wine.

"Well, now I'm really glad we didn't invite Sophie," Ian remarked.

I took another drink. "Have you ever had doubts about working at Harvester?"

He looked at me thoughtfully. "You know, Narin, sometimes I wish I could go to sleep and never wake up. But that's not fitting for people our age. We can't just drop dead without people thinking we're responsible in some way: alcohol and drug abuse, HIV, or just an inability to handle pressure. There's no reason to give your enemies that satisfaction. We must finish what we started. That's the only choice at this point. That's why I do what I do."

I understood what he meant.

"Do you like hiking in the woods?" Ian asked. I wasn't sure where he was going with this.

"Sometimes," I said, "Sophie and I have gone on a couple of hikes."

"You always start a hike full of hope," Ian said. He kept slicing the remaining bits of his steak. "You think you will see all sorts of amazing things. But after a while, you get tired and want it to end. I'm at that point now. I'm tired. I want it to end."

I nodded in agreement. "I heard you're getting a new assistant," I said.

"Dealing with all these legal matters just exhausts me. I'm lucky to have good staff working for me." He sighed. "Most people don't realize that any legal issue, any lawsuit, takes thousands of hours of work from people who might prefer to do something more productive and fulfilling with their lives than what the legal field calls 'discovery.'"

"It's going to be even more work when you and your dad take over the whole mess," I said. *Oh, fuck. I need to shut up. What was I saying?*

Ian nodded and smiled. I wasn't sure if he understood what I'd just said. I was slurring my words at that point.

I changed the subject. "What the hell is wrong with Maru anyway?" I asked. "Why does he always act like such a prick?"

Ian shook his head. "Well, to quote the great poet, 'No man is an island, but some do resemble peninsulas.'"

"He's always scowling at me," I said. "I'm pretty sure he hates me. And I have no idea why."

"It's nothing personal," Ian replied. "He's a hateful person."

"But why?"

Ian looked up at me but didn't say anything. He finished his steak, and since he wasn't interested in talking about it, I let it go.

"So, who else is on the leadership team at Harvester?" I asked.

Ian nodded. "Well, Sanjay Chandra is the head honcho, CEO, and Chairman of the Board. There's a CFO named Joseph Klein. You haven't seen him because he's been traveling in Europe. There's also a Vice President of Marketing and Global Outreach, Paula Oversby. She was once quite charming, bright, and hardworking. She seemed like a perfect fit at the time. I recommended her for that position. But ever since her promotion, she's become like the female equivalent of Maru. No one can stand her now. She's become repulsive, like fine champagne that's lost its bubbles. We'll be eliminating her position during the restructuring."

Ian poured himself another glass of wine. "What else? Maru is the VP of Drug Development. There's the VP of Corporate Affairs, and an open position for VP of Operations. We should combine compliance, risk management, and legal into a single VP-level role. If that happens, then I'll take that position myself." Ian examined his knife in the dim light. Tiny strands of meat clung to the serrated edge. "Maru had every opportunity to do something good with his life. And yet, he turned into a horrible person. It's unfortunate."

I stared at Ian, overwhelmed by the thought that he was my best friend in the entire world. And yet—

"We've become friends so quickly, and yet I feel like I barely know you," I remarked.

"What do you want to know?" he asked.

"I want to know about your father," I said casually.

He looked at me with a confused expression. "My father?"

"Sure. You know my father," I explained. "I want to know about yours. Didn't you say he was on the Board of Directors at Harvester?"

He paused to collect his thoughts. "Well, let me tell you a story then," he said.

Thirty-two

Ian began, "Harold Blair made his money in the insurance industry during the mid-1960s. This was a time when insurance companies often canceled policies to avoid large payouts. His talent was in mastering questionable practices that skirted the line between legal and illegal, and he became quite wealthy as a result."

I nodded as I watched Ian's mouth shape his words.

Ian took a drink and wiped his mouth. "He divorced his first wife and married my mother. She was a platinum blonde who was twenty years younger than him."

"What's her name?" I asked.

"Angela," he replied, not looking at me. "She had ties to St. Louis, so she forced him to come here. I was born a few months after they got married." Ian ran his finger along the rim of his wine glass as he spoke.

"We lived in a large mansion hidden behind a tall fence and a long driveway, almost like a fortress. Those walls kept the bad people out but also kept us trapped inside. My father was often away, flying to California whenever he could. As for my mother—" His eyes briefly lost focus as he squinted to see better. "I just don't remember her very well." He looked in my direction but didn't seem to really be looking at me.

"What happened to her?" I asked. "Did she die?"

Ian kept staring right through me. "I do remember her tantrums," Ian said. "She was always screaming and striking out." He shrugged. "There must have been good times: family outings, backyard barbecues, and pool parties if photos are to be believed. I don't remember that part. I was young when she died."

"I'm so sorry," I said. "How did she die?"

"It's like trying to remember a dream," he mused.

"Was she ill?" I asked. I was tempted to follow Ian's gaze and look behind me to see what he was seeing.

Ian continued, "She was ill. It was hard to tell just by looking at her. She had long tan limbs, a supermodel face, and flowing blonde hair, but her hands were always cold. I do remember that. I remember those freezing hands of hers." He picked up and swirled his glass. The wine slowly trickled down. "Those hands were always busy, uncovering evidence of offenses against her, rummaging through my father's belongings, pulling pockets inside out, reading his private notes. She always managed to find something. Then she would lock herself away and cry for days."

I nodded sympathetically, feeling sorry that I made Ian bring up those painful memories.

"I've always wondered why she didn't just leave him," Ian said. "She had her own money. Maybe she enjoyed the drama of it all. Maybe it was because of me." Ian paused for a moment. "I don't know."

His gaze shifted back to me. "To answer your question, she died of an overdose when I was ten. I was told it was sleeping pills. I'm the one who found her like that." He paused again, as if recalling something, and took another drink. "She was so beautiful. She looked more peaceful then. She'd always been restless, but now she was at peace, like an angel."

"I'm really sorry."

Ian smiled. "I'm sorry. You asked me about my father, and I'm boring you with old stories."

"I'm not bored," I said lamely.

"My dad and Sanjay Chandra were college friends. When Morey and Sanjay started Harvester twenty years ago, they asked my dad to join the Board of Directors."

I had a lot of questions, but I couldn't think of anything to say.

Ian said, "The whole episode with my mom left me quite traumatized."

"You've done very well for yourself," I said, "despite your hardships."

Ian smiled. "There are some benefits to being a basket case as a kid. That's how I met Floyd. We both got a lot of therapy when we were kids. I helped him get his job at Harvester."

"Does Floyd work at Harvester?" I asked. I hadn't seen Floyd there.

"He works in maintenance," Ian said. "If I hadn't met him in therapy, we would never have become poker buddies."

I sat quietly, gazing into Ian's sky-blue eyes. You might think my instinct would be to go over, hug him, and show my appreciation for his willingness to share such painful memories with me. But that's not what my gut was telling me. Instead, I was eager to seize the opportunity to strike while Ian was vulnerable.

I understand that's how bullies think; they target those who show even the slightest sign of weakness. But I didn't care. There were things I needed to say to Ian, and this was my chance. He might be good at hiding his secrets, but this could be my only opportunity to extract some promises from him. I had to act now and act decisively. I couldn't let this chance slip away.

Ian looked up to meet my eyes. "You're tired. Perhaps it's time to go."

"It's the wine." I finished my glass. "Wine always makes me sleepy. But—" I watched his eyes. As soon as he turned away, I seized my chance. "But it also loosens the tongue. Isn't that how the old saying goes? It makes you say what's on your mind."

Ian looked at me with curiosity. "Okay, I'll bite. Say what's on your mind."

"There are two things I want to talk to you about today." I felt a burning fire inside me. "Ever since we met, you've been telling me what you want for me, and I've been going along with your game. I've agreed to all your plans. I've done everything you asked. But I want you to know exactly what I want from our relationship."

Ian leaned back but stayed silent.

"I want the DMTA portfolio," I spat. "I want full control of the project. I want Morey out and me in charge."

Ian groaned, "Now we're back to killing Morey again."

"You don't have to kill him." I grabbed and squeezed Ian's arm. "If you ever get the chance, I want you to push him out of the way and make me in charge." I don't know why I was saying all this; maybe because I was drunk. But deep down, that's what I wanted. If he and Harold succeeded in their coup, then Ian would have the power to make me head of Neuroscience.

"May I ask why?" He scowled at me.

"Morey's brilliant, but I don't trust him," I said. "I don't trust him at all. He could screw me over, and there'd be nothing I could do about it."

Ian nodded. He understood that Morey couldn't be trusted.

I continued, "Something of great importance will come from our work with DMTA at Harvester, something with far-reaching effects beyond this project we're doing for the government or whoever. I want my name to be linked to it, not Morey's. I need to be the one in charge."

Ian shrugged and pulled his arm away. "If I am ever blessed with the opportunity to provide that service to you, I will do my best to ensure that you receive full credit for all the work on DMTA at Harvester." He took a deep breath. "The truth is, I probably won't have to do anything. Morey's a drunk and a gambler. He's deep in debt. He's on a fast and painful road to self-destruction. I wouldn't be surprised if he disappeared one day and never came back. What else?"

"Sophie's mine," I shouted. I didn't mean to yell, but I did. "It's obvious you're obsessed with her. You wouldn't know everything about her — where she is, or who she's with — if you weren't obsessed. You may even be in love with her, for all I know. And I know enough about girls to see that I wouldn't stand a chance if you went after her."

"You flatter me," Ian laughed.

"I'm not kidding about this!" I yelled, raising my fork fiercely.

Ian looked annoyed. "You just told me you weren't sure things would work out."

"That doesn't matter," I said. "You can have any other girl in the world. Just stay away from Sophie. Those are my conditions. DMTA's mine. Sophie's mine. If you want me as your obedient little sidekick, that's my price."

Ian shot me a disgusted look. "If I were in love with Sophie, why would I have introduced you to her?"

"Maybe you thought that if the two of us became dependent on you and I suddenly died, she'd fall right into your hands."

Ian laughed. "Is that what you think? That was my brilliant plan all along?" I didn't like him laughing at me.

"Is that why you introduced Ravi to Sophie?" I snapped. "But you didn't realize that the accident would almost kill Sophie, too. And so, you waited for me to show up so you could do the whole thing again."

He shot me a venomous look. "If that's what you think of me, then you can just go fuck yourself."

Ian waved to Jill, who approached us and handed him the bill. He checked it and gave her his credit card. She processed the payment and returned the card to him. He signed the screen, and she smiled when she saw the tip he left.

Ian asked her, "Are you available?"

"For you, always," she said with a smile.

"Are you ready to go?"

"Whenever you are."

Ian stood up and draped his arm around her. "Jill and I are heading out. See you at the office," he said.

"Thank you for dinner," I whispered. *What else could I say?*

Ian and Jill reached the door, but Ian signaled for Jill to wait. He staggered back to me. "Just what do you think is going to happen, you little shit?" Ian leaned on the table with both hands. "You think you're destined to be successful just because you're smart? Are you and Sophie going to live in a beautiful house and go on ski vacations with your gorgeous kids every winter? Is that what you picture in your mind's eye? They say every American sees themselves as a billionaire temporarily down on their luck. That's one thing you and Maru both have in common. You're both fucking Americans, through and through." He pushed himself away and walked out with Jill.

All I could think about was how I had fucked that up. I should never have gotten drunk. Ian hated me now. I needed to fix things immediately. I needed Ian way more than he needed me. And I had to get home safely, even though I couldn't see straight. *What the hell had I done?*

Thirty-three

The next day, I went directly to the ninth floor to apologize. I knocked on Ian's door, and he told me to come in. He was, as always, feeding his bird. Before I could say anything, he said, "I'm sorry for abandoning you last night. I hope you had no trouble getting home."

I didn't know what to say. "I'm sorry for being such an asshole," I stammered. "I didn't mean anything I said last night. I was drunk, that's all."

He signaled for me to stop. "Drunken conversations between friends sometimes get heated," he replied. "Ravi and I had quite a few shouting matches back in the day. Please, sit down."

I took a seat.

"I've given it some thought," he said, "and your requests are completely acceptable. DMTA and Sophie are now and forever your domain. Do whatever you wish with them."

I wanted to ask him about specifics, but I decided not to. Did I believe him? I guess I had to. I poured a cup of coffee for myself and one for him. He came over, sat next to me, picked up his cup, and took a sip. "Just the way I like it," he said warmly.

He looked incredibly dapper that morning. "Listen, Ian," I started. "You're my best friend. You can do no wrong as far as I'm concerned." I heard my voice crack.

"You're far too kind," Ian said. "Based on what you told me last night, our project will end soon. Then, we can move on to bigger and better things. We've got a plan—our plan. Let's stick with it."

"But what about Morey?" I asked. I wasn't even sure what I was asking. "How can we be sure Morey won't interfere with me getting credit for our DMTA work at Harvester? What about Morey blowing up when he finds out I'm dating Sophie?"

"Don't worry," Ian said as he leaned in closer. "Has anyone ever told you that you worry too much?"

The bird screeched loudly.

Ian looked at her. "You're fussy today. Is anything wrong?"

The bird watched us while we sipped coffee but didn't say a word.

"So, how did things turn out last night with Jill?" I asked, eager to hear the details.

"Well," Ian said, laughing, "we went back to her place right after the restaurant." He looked like a kid sharing a juicy story, his cheeks slightly flushed. "Jill led me into the bedroom and sat me on her bed facing the bathroom. She went into the bathroom but didn't close the door. As she stood at the vanity looking into the mirror, she pulled off her top and stretched to hang it on a hook on the wall. She's so skinny that the lines of her ribs were visible under her skin when she did this."

I pictured her slender body stretched out like that, the purple streaks in her dark hair shimmering in the light, and her dark animal eyes gazing into the mirror.

Ian continued, "She stood there topless, looking in the mirror for quite some time, examining her face and shaking out her hair. Then she turned her body toward me but kept staring at herself in the mirror. With a quick motion, she slipped off the rest of her clothes, picked them up, and hung them up. She had a— a Brazilian."

He looked at me curiously, and I was sure he was thinking about asking if Sophie did, too, but he decided against it.

"Finally, she turned to me, and our eyes met," he said. "She raised her arms as if to ask, 'Do you like what you see?'"

"And did you?" My mouth was dry, and my heart was pounding. I was painfully aroused.

"I did," Ian said with a laugh. "Very much so."

I shook my head. "Well, you had a better night than I did," I said.

After coffee, I went to the lab and saw Ahmed waiting for me in the common area. He signaled for me to come over and talk, as if he had something important to share.

"There's no hotter topic than Red Sky addiction," he stated. "Let's write a manuscript with our recent work and send it to 'Nature Translational Medicine'. What do you think of that?"

"Did you ask Morey?"

"I did," he said. "Morey told me to write a first draft and have you and JoAnn review it. He, of course, must approve the final draft."

"Yeah," I said. "Sounds good."

I wasn't really excited about being the third author, but I had to go along with it for now.

That Friday at the poker game, the main topic of conversation was Maru's poorly received email.

"All Sanjay wanted was for him to revise the mission statement," Chris exclaimed. "It's just a stupid mission statement. It's just a lot of happy, sappy talk about saving the world that no one cares about or reads. What kind of idiot could mess that up?"

"Well," Ian said, "our friend, Maru."

"All the division heads are furious," Carlos said. "Some are threatening to leave. Sanjay had better fix this quickly. This is not something that can wait until Monday."

I hadn't received the email, so I wasn't aware of what Maru said that was so offensive.

The conversation then shifted to the July 4th celebration at Ian's house. Ian invited everyone to come. Alex complained that his van was still in the shop and asked if anyone could give him a ride to Ian's party. I volunteered. I'm not sure why. I didn't like Alex, but I figured he couldn't be all bad since Sophie had dated him.

For the final hand of the night, the pot exceeded four hundred dollars. Only Carlos and Ian remained in the game. All eyes were on Ian, who often swelled the pot with poorly executed bluffs. However, this time felt different, and Carlos's grimace showed he was deep in thought, making complex calculations.

"Until the moment you show your cards, anything is possible." Ian held his cards close to his chest. "But that doesn't change what's in your hand. Only someone with perfect knowledge knows the outcome."

Carlos looked at Ian. "But which of us mortals has perfect knowledge?"

Ian smiled and let out a hearty laugh. He set down his cards to reveal his bluff. Carlos wiped his forehead and collected his winnings.

On my drive home from the game, I thought about the Desert Prince. I hadn't seen him in a while and wondered if I had just imagined the whole thing. Or maybe he was done with me after I defeated Monkey. I wasn't sure. I'd been slacking off on his prescribed training regimen, so I spent Saturday catching up on a week's worth of missed training.

I visited Sophie's house on Sunday morning. She had been overwhelmed that week, trying to finish a bunch of orders. If one of her paintings sold on Etsy or eBay and she received another order for the same painting, she would quickly make a copy and ship it. She called these 'duplicates'. I'm not sure if this was ethical. I imagine everyone who ordered a painting thought they were getting a unique piece, but I didn't fault her for doing what she needed to make money. I admired her hustle.

To my surprise, she wasn't painting when I arrived. She was curled up on the couch, wearing an oversized sweatshirt covered in paint and some ragged old sweatpants. Her eyes were bloodshot, and I could tell she'd been drinking — not just wine, but her favorite gin. She'd told me she often dreamed of lying on a white sandy beach in a bikini, feeling the warm ocean breeze on her skin, drinking an ice-cold gin and tonic. Usually, she knew better than to drink hard liquor because she shared her father's tendency. But when she felt depressed, she couldn't resist. I wondered how much she had drunk that morning.

I sat next to her and took her hand.

"I spent the whole morning trying to make an instructional painting video that I was going to post online," she said in a small, mournful voice. "It was a complete disaster. The sound was terrible. The lighting was all wrong." Tears rolled down her face. I reached out and held her hand.

All I could think was, *Girl, you're not going to get anyone to watch your videos dressed like that! You have so much natural beauty, and you need to show it off to build an online following. If people see a beautiful girl, they'll overlook poor sound and lighting.* Of course, I didn't say that to her; I knew better than that. She was distraught.

Her tears turned into sobs as she gripped my hand even tighter.

I gently asked her, "Do you want to be alone?"

"Just give me a moment," she said. "I'm sorry."

"Don't be sorry!" I said emphatically. "I only wish I could be more helpful. I wish that I could maybe do something—something—"

She sat up, and I wrapped my arms around her.

"I love you, Sophie," I said.

She stopped crying and stared at me. "I love you too, Narin," she said. "More than the whole world." She wiped away her tears.

I wasn't sure if I believed her. She might have just been caught up in the moment. After all, as far as she knew, I was still just a lab grunt. She would have a genuine reason to love me someday when I became a renowned scientist. I needed to push Morey harder, work harder, and start on the DMTA project that could change the world.

"I need to talk to you about something," she said. "Something important."

I waited. She had my full attention.

"It's about the accident," she said. "I haven't told anyone, but I need to tell you. And I need you to believe me."

"Of course," I replied.

"It's like a bad dream, that accident. It happened so suddenly." Her face twisted as she remembered the events.

I held her close. "Tell me," I said.

"He used to use Red Sky," she said. She stated it so matter-of-factly that it didn't register. I already knew Ravi was a Red Skyer, but I was still surprised to hear it from her.

"What did you say?"

"He said it helped him relax." She pulled away from me. "It helped him be more creative. I didn't mind it since he was different, gentler, and more caring when he took it. It didn't seem to affect him much otherwise. He'd pop them all the time. He never hallucinated or walked around aimlessly like Red Skyers do. He never mumbled to himself. He could carry on conversations about philosophy, art, literature, or anything—no problem at all. You'd never know."

The realization was sinking in. "So—Ravi was using Red Sky when you had the accident?"

She stiffened.

"He was!" I shouted, my fury boiling over. "He was under the influence. He caused the accident that killed him and nearly killed you. That's what happened, isn't it?"

"That's not what happened, Narin," she said, but she didn't sound very convincing.

"That's exactly what happened! Why are you defending him? He almost killed you! You don't have to defend him anymore. He's dead and gone."

"I'm not defending him, Narin," she said. Tears streamed down her face again. "I'm telling you that's not what happened. He could have taken a whole bottle of those stupid pills, and it wouldn't have mattered. They didn't affect him." She rubbed her forehead, exhaled deeply, and wiped tears from her eyes. "What I'm trying to tell you is that he took one stupid pill, and his whole body started shaking. He completely lost control. We'd swerved into the other lane before I realized what was happening. It was all over in a matter of seconds."

I listened to her every word and watched her every move with an intensity I doubt I'll ever be able to match again. "So, what exactly do you think happened then?"

"I think someone poisoned him, whoever was giving him Red Sky."

I tried to think. *What was she saying? Did someone murder Ravi?*

"Who was that?" I asked. "Who was his supplier?"

She shook her head and shrugged.

"Did you tell the police?"

"I was in the hospital for four weeks, and no, I didn't tell anyone. You're the first person I've ever shared this with."

"Did he have any enemies?"

"We all have enemies," she said. "The problem is figuring out who they are."

I had no idea what to make of this strange story. Had Ian given him tainted Red Sky? I thought back to our drunken conversation from the night before. *Was I right? Did Ian kill Ravi so he could have Sophie? Did that even make sense?* Things were getting so tangled in my mind that I couldn't think clearly.

But I realized one thing: I had to be careful. There would be no more coffee in Ian's office, and I was going to carry the knife with me wherever I went. Whatever danger the Desert Prince had prepared me for was now nearby, and I had to be ready for anything.

I was lost in my own thoughts during dinner with the professor. As we were leaving, he asked me to stay for a moment so we could talk. Sophie and I looked at each other, both of us unsure of what this was about. Sophie went out, and I sat down. The professor cleared a spot on the couch and sat across from me. He looked troubled, as if he didn't know where to start. Finally, he said, "I understand that you are working for Harvester Pharmaceuticals."

I nodded.

"You study the effect of savycha on the brain."

I wasn't sure whether that was a question, so I nodded once more.

He leaned forward and looked at me. His tremor was worse than I had ever seen.

"Sophia told me something interesting the other day. She thought it might have been an inside joke between you and your housemate. She mentioned you were discussing the development of a truth serum."

My heart was pounding. *How did he figure it out?*

He continued, "It reminded me of a story I heard during my travels in South America about how, in the 1600s, those accused of witchcraft were force-fed savycha by the parish priest to obtain a confession."

He kept staring at me over his glasses, and I felt a mix of fear and anger building inside. I needed to say something. What should I say? Should I laugh it off? Or warn him that he had crossed into dangerous territory? Operation Juggernaut was top secret.

"Have you talked to your father about your work?" he asked.

This pissed me off. "You may not have noticed," I said, "I'm thirty years old. I don't need my father's permission to do anything."

He shook his head. "Your father is a very wise man," he replied. "You may find his guidance useful."

I shouted, "I have never asked my father for advice, and I never will."

His eyes darted around, and he seemed less sure of himself. Maybe he saw something in me that scared him. If he told my father what I was doing at Harvester and my father mentioned it to any of his friends, the rumor would spread quickly through the Indian community. The people who commissioned Operation Juggernaut wouldn't be happy if this became public, and the leak would definitely be traced back to me. I had to stop this before it got out of hand. Sanquel, Sophie, the professor, and maybe even Peter had become threats to my survival.

The professor looked toward the dining area as if searching for an escape route. I couldn't help but smile at the thought of him trying to run from me. I'd take him down before he could take two steps and smother him with a pillow. Sophie would find him like that the next time she came over.

What was I waiting for? I should close the gap between the decision and the action, just as the Desert Prince had said. It would be so easy that it would be laughable. I held my position, glaring at the professor with murder in my heart.

I was considering killing Professor Goldblum. Part of me was shocked at what I had become.

"Your father will return soon," he whispered. "You should talk to him about what's bothering you before it's too late."

I got up and walked out.

I went to Sophie's house, and she started bombarding me with questions. "What was that about? Is he sick? I know he doesn't want me to worry, but he should go to the doctor if he's not feeling well."

I could barely look at her. She betrayed me by gossiping about things that weren't her concern.

"What's wrong, Narin?" she asked, almost in tears.

I left without saying another word.

Thirty-four

It was disappointing because I was really looking forward to the sex that night. I'd been staying over with Sophie quite often by this point, and we were having a lot of sex. In fact, I was getting pretty good at it.

I went home and got ready for bed. I had cooled off. I was still upset with Professor Goldblum and Sanquel, but how could I be mad at Sophie? She had no idea what she was doing, revealing my secrets like that.

I had just gotten into bed when Deepa called me. She told me our parents were heading back home and wanted me to come to Chicago. They were hosting Rajeev for dinner on Friday and told me to be there. "I'm looking forward to it," I said, even though I wasn't, and hung up.

Monday marked the official end of our involvement in Operation Juggernaut. The data looked perfect. Now, tests on enemy combatants would show if we had succeeded. I wasn't sure how much JoAnn, Ahmed, and the techs understood about the project's nature, but everyone was in good spirits. By late afternoon, Morey called me into his office. His speech was so slurred it sounded like he'd had a stroke. When I arrived, he motioned for me to sit down. He wasn't aware that I knew about Operation Juggernaut, so I stayed alert to avoid revealing anything.

Morey looked at me thoughtfully, almost affectionately. "A hundred years ago, St. Louis was poised to become an international city with thriving industry, culture, and art, hosting the World's Fair and the Olympic

Games in 1904. People converged on this burgeoning river city, called the Gateway to the West. At that time, Chicago was a scrappy Midwestern town mainly known for its organized crime. But somewhere along the way, something went wrong. St. Louis ultimately became a mediocre, backwater town of little importance. Most Americans wouldn't even be able to identify it on a map. Chicago, by comparison, has taken on the mantle of the metropolis of the Midwest. You see, Narin, even simple, seemingly insignificant events can knock a city or a person off the path to greatness. That's been the story of my life." He rubbed his face. "Bad decisions and bad luck have ruined things for me."

"Did you want something from me, Morey?" I asked.

"I wanted to congratulate you," Morey said. "If it weren't for you, we wouldn't have finished our little project. Now, we have a head start on developing a way to tame Red Sky. We're sitting on a goldmine." Morey nodded. "The money will start pouring in soon. Right away, there's a hundred-thousand-dollar bonus for me and thirty thousand for each of you."

I was surprised to hear that. They hadn't even tested it on humans, and no one mentioned anything about a bonus.

"That is good news," I said.

Morey pulled a flask from his jacket and took a drink.

A minute passed before Morey spoke again. "I think we need to get our pharmacokinetic data out there. We should aim high, maybe even *Nature* or *Science*. Ahmed should be the first author since he's been here the longest, and JoAnn the second, but you can be third."

I nodded.

Morey continued, "I talked to Pharma, lit a fire under their ass. With Sanjay's backing, we'll get what we need soon. We need to tweak your Type II receptor blocker to improve its oral bioavailability. We should have everything ready in the next six months. Hell, maybe if we combine it with an off-the-shelf antidepressant, we can prevent suicides entirely. A lot of good could come from all this work."

Morey rubbed his brow. "We might need to bring in more people from our department for this project. Chandra doesn't want me to hire anyone new, but we can reassign some staff and redistribute the workload. Is there anyone you'd want on your team?"

I didn't say anything. I wanted to end this conversation smoothly, but I had no idea how to do it.

Morey looked at Sophie's picture. He picked it up and studied it for a long time.

"You're smart, capable, and level-headed," he told me. "You remind me of myself when I was younger."

What was he talking about? Had he heard about Sophie and me? Was this his way of giving me his blessing?

"She never told me she was attending art school," Morey said wistfully. "I didn't discover her intentions until her junior year, by accident. She knows how to keep people in the dark about her intentions."

I was tempted to say, 'I wonder where she learned that from,' but I kept quiet.

Morey set down the picture. "My lab-mates were jealous of me also," he said.

"You think Ahmed and JoAnn are jealous? Of me?" I was stunned to hear him say that.

He nodded and took another swig from his flask. "I don't blame them," he said. "You're a man with a bright future ahead of you."

I wasn't sure how to respond to that.

"I'm not going to be around forever," Morey continued. "You stick with me, and I'll make sure you get credit for all your hard work. I've got Sanjay's ear. I'll let him know how talented you are. I'll be the best advocate you've ever known. Of course, that doesn't mean much considering the shit-hole bosses you've dealt with in the past."

We sat in silence for a long time. I wasn't sure if he was serious or not. He might not even remember this conversation in the morning. I was tempted to use this moment to reveal what he might have heard whispered in the rumor mill: that I was in a serious relationship with his daughter. Maybe he was signaling that it was okay for me to tell him. Perhaps this drunken nonsense was all about that. Maybe he'd say that if I loved Sophie and she loved me, we should get married right away, and I should start calling him 'Daddy' from now on.

"Thank you for your kind words," I said. "You don't realize how much that means to me."

After I got home, I called Sophie. I should have told her about that strange conversation with Morey. In hindsight, that's what we should have talked about and had a good laugh over. Instead, I told her I was going to Ian's Fourth of July party the next day. When she didn't reply, I added that I'd be taking Friday off work to go back to Chicago and visit my parents over the weekend. "They'll finally be home from India," I explained. She was, as always, busy doing something and didn't seem to be paying any attention to what I was saying.

"Well, I guess David and I will have to celebrate my birthday on our own," she said at last. By David, she meant Professor Goldblum.

"It's your birthday?" I asked. "You didn't tell me that."

"You never asked," she shot back. "It's this Friday."

"Well, I'm sorry." I tried to think of a way to avoid going to Chicago, but I knew I had to be there. I didn't have a choice. Deepa would be furious if I tried to back out, but I didn't want to upset Sophie either. I was stuck, unsure of what to do.

"That's fine," she replied coldly. "You go and see your family."

"Why are you upset?" I honestly didn't know. "I'll be back by Sunday evening. I'll bring a cake to the professor's house. If you're nice to me, I'll get you a present from Chicago."

"I was just thinking you haven't told me much about your parents," she said sharply.

"There's not much to tell," I said. I should have seen where this was heading.

"Have you told them about me?"

The question caught me off guard. "They've been in India," I responded.

"Are you going to tell them when you see them?" This set off alarm bells in my head. I had to be careful about what I said. "I don't talk to them about all my friends, Sophie," I said.

A painful silence followed.

"What I mean is, we don't have that kind of relationship where we share every little detail about our personal lives," I blundered on. "I've got to take things slowly, lay the groundwork, so they don't get upset."

"Upset?" she yelled. "Why the hell would they be upset?"

I sighed. I absolutely hated it when people yelled at me. I'd gotten myself into deep shit, but strangely, I didn't care that much. I was glad I'd hit a nerve. It was just punishment for blabbing to Goldblum about a truth serum. Besides, trying to reason with her was a losing battle, especially when she was so upset.

"I guess I just don't understand," she said, her voice low and threatening. "It's tough being an outsider. Well, I know how busy you are," she said sarcastically. "I'm busy, too. I need to get back to work." She hung up on me. I expected her to call back, but she didn't.

"Well, at least I talk to my parents!" I shouted into my empty room. "How can you be critical of me when you abandoned your poor dad in his time of need? You're nothing but a spoiled, stupid, damaged little girl. I should have known this would never work out!"

I spent the night thinking about all the horrible things I could say to her, how she'd react, and then tell me that she was bored to death with me. She'd criticize me for lacking the adventurous spirit of Alex and the talent and charisma of Ravi. I definitely wasn't as handsome and charming as Ian. I was nothing but a born loser. She'd admit that she didn't care about me, and if she could, she'd wish me dead if it would bring back Ravi.

The more I thought about it, the more I realized she didn't love me. She never gave me her full attention. She was always working, only responding in disjointed pieces while she painted and did other nonsense. We had nothing in common and no real connection. I was just a distraction for her, a plaything. Now, it was all clear to me.

I replayed her angry words, both real and imagined, and soon it became difficult to distinguish between them. I spent the night caught in waves of self-pity. I didn't need her; if anything, she needed me. She didn't even realize all the things I could do for her. She'd regret ending things with me. She'd regret how she mistreated me.

By morning, I desperately wanted to call Sophie, sincerely apologize, and plead for her forgiveness. I'd tell her that she was right and I was wrong. I'd ask her for another chance.

That was Tuesday, July 4th. What a stupid day for a holiday! I had to pick up Alex and go to Ian's that evening. Being off work only made things worse. I stared at her picture on my phone for hours, but I couldn't make the call. I wasn't sure what was holding me back. I kept thinking about what Ian had said, about how she'd grow bored with me if I weren't careful. He was right. She was already tired of me, and this was her way of ending things.

She had to admit that I was kind, loving, and willing to do anything for her. I wasn't a cool adventure junkie like Alex, and I sure as hell wasn't a real junkie like Ravi. If those were the kinds of guys she fell for, then I wasn't the right one for her. What did she say about Ravi? He was gentler and more caring when he took Red Sky. He was probably a callous, self-absorbed asshole who abused and manipulated her. Maybe that's what she wanted, but I just couldn't be that person.

She must have enjoyed being showcased by Ravi to all his fancy friends in their exclusive New York and London salons. I definitely couldn't offer her that. I couldn't get over her agreeing to marry someone so much older than her. What was she thinking? My head started throbbing. It felt as if it was trying to give birth to a squirming bundle of snakes. I decided I wouldn't think about Sophie until I returned from Chicago.

Thirty-Five

When I picked up Alex from his parents' house in Maplewood, the sky was darkening. He was dressed in torn blue jeans and a black T-shirt.

"I've been living out of my van since I left school," Alex said. "It sucks not having it. Thanks for the ride."

We drove in silence, listening to music.

"Nice wheels," Alex said.

I didn't respond.

Alex cleared his throat. "I heard you're dating Sophie."

"I am," I said.

"How's she doing?"

I shrugged.

"You probably know the story already," Alex said. "We were planning to go away together. She wanted me to rescue her from her boring life. That's what she told me. So, we decided to head west to Yosemite and become park rangers. It was her idea, not mine. She said she'd always wanted to go west, to escape from St. Louis. I planned it for weeks and had everything lined up. But then—she changed her mind."

I sped up to merge onto the highway.

"I had no idea about Professor Reddy," Alex said. "I'm not mad about it anymore. I was then. But not anymore. I went to California alone, and it all turned out fine. You feel freer, less oppressed out in the wilderness than when you're kicking around here." He sighed, pulled out a cigarette, but then put it back in his pocket. "I honestly thought she'd come out there eventually. After all, it was her idea. But I guess she'd moved on. She's pretty good at moving on."

"Is that right?"

"Independence is a good thing," he added quickly. "We all need to make our own lives, but most don't have the balls for it. It takes courage to do your own thing, to blaze your own trail. It's much easier to hitch yourself to someone else and let them pull the load." Alex pulled out the cigarette again, held it for a moment, and tucked it back into his pocket. "She's not a bad person, Sophie," Alex added. "She's not like Morey. Sure, she drinks, but she's not a lush like her dad."

The lingering smell of cigarettes on his clothes finally affected me. I opened the window, and Alex did the same.

We drove silently for several minutes.

Alex chuckled. "The one thing that used to drive me crazy was that when she worked, she'd talk with you, but you could tell from her responses she wasn't listening. It's like her mind was somewhere else. I guess she can't devote enough attention to whatever you tell her when she's in the zone."

"She still does that," I said solemnly.

Alex laughed. "Let her know I'm here if she ever wants to reconnect. She just needs to say the word. I wish we'd kept in touch. I wish we were still friends. She's a great person. I only wish—" His voice trailed off.

I saw the signs for St. Charles and took the exit ramp.

"Do you ever catch her talking to herself?" Alex asked.

I shook my head.

"It's only when she's sloshed, usually on gin," he said. "Now that's a girl who likes her gin."

"I've never seen her drink," I lied.

Alex once again took out the cigarette and held it lightly in his hand.

Ian's house soon came into view. It was a glass and brick monstrosity perched high on a hill. The setting sun reflected painfully off the glass panels into my eyes, threatening to give me a headache. Bright, multi-colored lights and loud bass music blasted in all directions. One road led down to a lake, while the other led to his house. We parked on the street and went inside.

I worked my way through the crowd and found Floyd sitting alone at a wooden table in the back corner of Ian's living room. He was looking out the window at the lake, where people were setting off fireworks from the dock. I sat down beside him.

"There are so many people here," I shouted over the music. "Are they all from Harvester?"

"Mostly," Floyd replied.

"Does Ian usually draw such a big crowd?" I asked.

"Don't know," Floyd said. "I don't go to Ian's parties. We have our family reunion on July 4th, but last year, something stupid happened, and I got banished."

"What happened?"

"I dropped my uncle's camera while taking a family photo, and it broke. My mom freaked out, started apologizing, and told everyone that the reason I was such a failure was that I never listened to her. It sucked."

"That sounds pretty awful."

Floyd shrugged. "I didn't get invited this year."

"Well, I'm glad you're here. It's nice to see someone I know."

I caught a quick glimpse of Maru, but he vanished into the crowd. I also noticed Susan, the receptionist from the lobby, wearing a short black dress and animatedly chatting with a group of older men in suits who surrounded her. I looked around for Morey, but I didn't see him.

Ian appeared out of nowhere. "It's good to see you both." He bowed with a flourish. "Come with me, Narin, and let me introduce you to some of my friends."

I followed Ian but seized the first opportunity to pull him aside into a corner. He shot me a curious look.

"I want to know something about Ravi," I said.

He nodded. "Sure. What?"

"Do you know who was giving him Red Sky?" I asked sharply.

He frowned. "Why do you ask?"

"He was a Red Skyer. Sophie told me. Were you the one who gave it to him? Yes, or no?"

"No, of course not," Ian said. "Maru was his supplier."

I tried to process this. "Maru was what?"

Ian snorted. "Maru's been a Red Skyer for years. He knew Ravi well. They're both what you call high-functioning addicts, even though it's a bit of a stretch to call Maru 'high-functioning'. Maru made a little side business selling to other well-to-do addicts. He has it shipped in from the Caymans in Harvester boxes, so it doesn't go through the normal customs process. It's all done on the sly, very illegal."

"Maru supplied Ravi with Red Sky?"

"Not just to Ravi," he said. "He still supplies a lot of the upper crust: doctors, lawyers, businessmen, politicians, and, as in Ravi's case, university professors. That's his usual clientele. I know his secrets, and they aren't pretty." A group of revelers bumped into Ian, and he smiled and laughed. "Come on, let's make the rounds."

Ian guided me through several rooms and introduced me to everyone we encountered. He appeared to know everyone, from the IT technician to the VPs. I looked around for Sanjay Chandra, but he wasn't there. It was odd to see these somewhat familiar women from Harvester, so reserved in their usual work clothes, now strutting around in short dresses, with flowing hair, shining eyes, glowing lips, and teeth flashing in eager smiles. They greeted me, though most seemed eager to talk with Ian.

"You're very popular," I remarked.

Ian laughed. "Flies are attracted to honey, but also to garbage."

"Is your father here?" I asked. "I'd like to meet him."

"I'm afraid not," Ian said.

It was a relief to return to Floyd. He was still in the same spot, gazing out the window. I sat beside him, still trying to process Ian's revelation: Maru supplied Red Sky to Ravi.

"Have you ever used Red Sky?" I asked Floyd.

"No, but I get why people do."

"To get high?"

"At least Red Skyers have the next hit to look forward to," Floyd replied. "Speaking of Red Sky, Maru came over earlier looking for you."

"What did he want?"

"He said you should check out the red room upstairs."

"What's that?"

"I don't know." Floyd shrugged. "He said it was the second door on the left upstairs." Floyd pointed toward the stairs.

I paused for a moment before jumping up and climbing the stairs.

I approached the second door on the left, grasped the copper doorknob, and knocked softly. "Maru, are you in there?" No response. I stepped inside and flipped the light switch. The room was dimly lit, with only three ceiling lights on the other side. All the walls were painted blood red. There was no furniture. An oriental rug covered most of the floor. No one else was there.

I closed the door and moved to the far end of the room. The three lights illuminated three paintings. I immediately recognized the subject, and my heart began pounding. All three depicted Sophie, though she seemed barely older than a child, maybe eighteen or nineteen. Her naked body was on full display in all three works. To the right, she sat on the ground, arms wrapped around her bent knees, with her eyes shining and lips forming a faint smile. In the center painting, her legs were tucked beneath her, revealing her slim, graceful back. She looked back over her left shoulder, giving a sultry glance at the viewer. The third painting showed her lying on her side, knees bent, her head resting on a velvet pillow. She arched her back slightly, making her small, round breasts stick out. Her eyes were half-closed, with an expression of bliss. She looked like she was about to fall asleep or perhaps was heavily sedated.

I examined the distinctive signature on each painting: "Ravi."

"And I thought you were into skinny brown boys, Ian," I muttered to the empty room. I stood frozen in front of the three paintings. I could barely bear to look at them, yet I couldn't look away. She was so achingly beautiful, but intense anger and regret washed over me, as if Sophie's innocence had been torn apart and sullied by Ravi. My Sophie. Not my Sophie.

Tears welled up in my eyes. I imagined ripping the paintings off the wall and setting them on fire. With effort, I turned away from the images, switched off the lights, and left the room.

Floyd was still in the same spot where I had left him. "Did you find it?" he mumbled.

"I did," I whispered. I watched the revelers talking and laughing, trying to distract myself. My eyes fixed on a young Indian girl in a tight dress as she moved around the room. She received hugs from those around her.

"She's cute," Floyd said.

"Who?"

"That girl you're staring at. Why don't you go talk to her?"

"Why don't you?" I replied irritably.

"Being social isn't my thing," Floyd replied.

I moved across the room and stood behind her while she talked to a group of girls. Her voice sounded very similar to my sister's. Their conversation continued without interruption as she took a drink from the passing server. When I couldn't take it anymore, I gently tapped her shoulder, and she turned around.

"Hi, I'm Narin Roy," I said. I braced myself for the 'why are you bothering me' look, but instead she smiled. Her lipstick shimmered under the bright overhead lights.

"I've heard of you," she said. "You're Ian's friend. Hey, let's take that sofa before someone else sits there. I've been on my feet all day and need a break." We squeezed onto the seat. She took off her heels and put them under the sofa.

"That's so much better," she said with a contented sigh. "So, how long have you known Ian?"

"We've been close for a long time," I said. "You could almost say we're best friends."

She seemed impressed by this.

"Do you work for Harvester?" I asked.

"I'm a drug rep." She gestured back at the group she'd been talking to. "All of us girls are. Jack's the only guy rep there." She pointed to a tall, handsome young man. She leaned in closer. "I heard from the guys here that Ian's in a serious relationship. Do you know anything about that?"

"I don't think so." I was confused by this. I was sure Ian had plenty of girls (and guys) whenever he wanted, but nothing serious. "Ian has never told me about a serious relationship," I said.

"And you would know?" she asked, looking at me expectantly.

"I imagine." I wanted to steer the conversation away from Ian and back to myself, but I never got the chance.

The music stopped.

"Everyone," Ian shouted. We looked over and saw him standing on the dining table. "Please, if I could have a word."

A hush settled over the room.

"I want to thank all of you for coming here today," he shouted. "As you know, this has been a very challenging time at Harvester, but it's also been exciting. Great things are happening at our beloved company. We owe a debt of gratitude to Sanjay Chandra for his steady hand guiding us through these troubled waters." There was a smattering of applause. "And now important projects are coming to fruition. I can only say: Hang on tight because it will be a bumpy but exhilarating ride. Although we may not be as big as some of our competitors, we undoubtedly have the most dedicated team of professionals working for us. The world will see very soon that Harvester is one unstoppable juggernaut." Another smattering of applause. "Okay, my friends, fireworks in ten minutes." Ian jumped down.

"I'm going to try to catch Ian before someone else does. It was nice meeting you." She left before I could ask her name.

What the hell was I trying to achieve? Would I forget about Sophie if I started seeing someone else? It felt pointless. The party was wearing me out, and I could feel a migraine beginning deep inside my head. I looked around the room but didn't see any sign of Alex and thought about leaving without him.

I waited for the rhythmic music to start again, but it didn't. I'd gotten used to its constant beat. Without it, something felt wrong. It was like the scene would fall apart without the dance music holding it together. I wasn't the only one who felt this way; others were beginning to get restless, too.

I wandered around the room, searching for those who once seemed so exciting: the pretty blonde girl in the bright blue dress and the petite redheaded girl who affectionately placed her hand over mine when she greeted me. It was a small gesture, but it sent a thrill through me. But all these lovely creatures had vanished. Instead, I was surrounded by people who repulsed me, laughing with their mouths wide open, voices loud and grating, and bodies on display in ridiculous costumes. I needed to get out of there. But not until I confronted Ian.

I didn't know what to say to him. I wanted to slap him across the face for disrespecting Sophie with those lewd paintings. Was it all just a ploy to fuck with me? I wanted to grab him by the throat and shake him until he told me what he wanted from me and Sophie. What kind of game was he playing? I was ready for a fight that would end all fights. I just wished I had my knife.

I finally found Ian in a corner talking with Chris Vender.

"What I mean to say," Chris said, his words emphasized with exaggerated gestures, "is that people our age need to start accepting our limitations. When you're younger, it's easy enough to dream of being a football star or an actor or something, but at our age, it's time to move on."

Ian nodded eagerly. The gesture reminded me of Monkey and how he always nodded at everything the Desert Prince said.

Chris continued, "I mean, there are some things I do well, right? Like my work at Harvester. I'm not saying I enjoy it or believe it's worth doing. But if I do the things I do well, nonetheless, then at least I can say that I'm doing something. Right?"

Ian kept nodding, but then he caught sight of me out of the corner of his eye.

"Excuse me, but I must get on with the show." Ian grabbed my elbow and pulled me out the front door.

"Follow me," he commanded.

"Ian, I need to talk to you." I hurried to catch up with him, but he was already far ahead and disappeared into the woods.

I followed him down a muddy trail. Soon, we reached the dock, and Ian untied a pontoon boat.

Come on, man. We gotta move. We're running late," Ian said breathlessly.

I climbed onto the boat, and Ian handed me a paddle. We rowed until we reached the middle of the lake.

"Untie your side," he commanded.

I took off the tarp covering the boat's hull.

"What's all this?" I asked, feeling totally confused about what was happening.

"Fireworks," Ian said. "It's the Fourth of July. Most of it is under computer control, but you need to make sure the pilot light stays lit and doesn't blow out."

A small flame flickered through a dirty black grate.

Ian shouted to the crowd on the balcony, "One minute!"

He grabbed my arm and pulled me down. "Lie down next to me," he said. "You're shivering, man. It's like a hundred degrees out here. Just relax. The fireworks are completely safe—or at least that's what the one-armed salesman told me."

I let out a sigh. "There's something I need to talk to you about."

Ian looked at me. "Go ahead, but make it quick. Once the fireworks start, I won't be able to hear anything."

I tried to gather my thoughts, but the words wouldn't come. Everything was a mess: Ian, Ravi, Sophie, Red Sky, the deadly car crash, poisoning, vulgar paintings, and fireworks. Finally, I asked, "What did you do that got you expelled from college?" I had no idea where that came from. I'd thought about it now and then, but it wasn't at the forefront of my mind.

"I stabbed someone," he said.

An electric wave rushed through me.

"It was stupid," Ian said. "I was drunk. It was outside a bar."

"Did he die?"

"She's alive and doing quite well thanks to the money she took from my father."

The boat began to rumble.

"It was all so idiotic," Ian said. "It's the stupidest thing I've ever done."

We lay in silence, waiting for the fireworks to begin.

"When I was in high school," I said, "I finally found the courage to ask this beautiful girl to dance with me. During the dance, she told me I danced like a wooden Indian."

Ian considered for a moment. "What did you say to that?"

"I told her she was right. She was so beautiful that she made me into a wooden Indian." I gave Ian a crooked smile. "A wooden Indian."

"I get it," Ian laughed. "Very smooth. No wonder the girls love you."

At that moment, I felt deep affection for Ian. I owed him everything. Without Ian, I would have to find a job outside of science. With my pharmacy background, I probably would end up stuck behind a Walgreens counter, dispensing oxycodone prescriptions. He had given me the chance to earn recognition for all my hard work. And he brought me Sophie. He single-handedly made my greatest desires possible. Ian was my best friend. I would follow him to the ends of the earth if that's where he was going.

Flames burst from several openings on the top of the boat. The boat swayed gently as fireworks shot into the sky and exploded.

"This is the best view you can get," Ian yelled, pointing to the bright colors glittering overhead. The crowd cheered, and ash drifted down onto the lake.

I'd forgiven Ian, but I wasn't sure if I could forgive Sophie. You might wonder what she'd done that needed my forgiveness. She'd made me insanely jealous, and it hurt like fucking hell.

Thirty-Six

On Wednesday, July 5th, Morey didn't come to work. JoAnn told me that he often went on unannounced benders and wouldn't show up for days. With our project finished, we were left in limbo. Ahmed spent his time writing our manuscript. JoAnn and I reviewed the experiments from last month to make sure we hadn't missed anything. Sanquel avoided me that evening, and I only found out why the next day.

Morey also didn't come to work on Thursday. Ahmed and I tried calling him several times, but the calls went directly to voicemail.

"That's very strange," Ahmed said. "He mentioned he would get a large bonus upon completing this project with H138. Perhaps he has flown the coop with his newfound wealth."

I shrugged. "I guess that means you're in charge now."

That evening, I found Sanquel in the living room waiting for me, sitting with the lights dimmed. He asked me to join him, so I did. He looked worried, but I didn't understand why. I sat down in a chair across from him to see what was happening.

"I—" Sanquel started, "I wanted to tell you that Sophie came over on Tuesday." He paused. Sanquel, who usually spoke eloquently and confidently, seemed at a loss for words. I didn't say anything; I just looked at him.

"She might not have realized you were going to your friend's party."

I kept staring, knowing whatever he was about to say would piss me off.

"What did she want?" I asked coldly.

He looked up at me like I'd broken a spell.

"She said she was exhausted from painting all day and wanted to take a break. She seemed quite disappointed when I told her the unfortunate news that you weren't home." He nodded as if to convince himself of that. "Of course, I invited her in and offered her some refreshments."

"That was very gentlemanly of you," I remarked.

"Fortunately, I had food ready: a simple meal of braised salmon."

"And wine?"

Sanquel considered. "A dry white wine, like Pinot Gris, would have paired better with this meal, but all we had was Chardonnay. But she was okay with that."

"I'm sure she was," I said. "That sounds lovely."

"I offered to call you to let you know she was here, but she said she didn't want to bother you."

I nodded.

"She wanted to know if Peter and I would be interested in going out with her to celebrate her birthday on Friday. Of course, I said the affair would be much more enjoyable if we waited for your return." He paused for a moment. "Peter is quite busy, as am I."

I thought that choosing the word 'affair' might have had some hidden meaning. I stayed silent even though I desperately wanted to know how this ended.

Sanquel continued, "As we ate, Sophie told me about her artistic and financial struggles. She said that she was increasing her sales and was starting to be able to save a small amount each month, but she still felt like she wasn't making much progress." Sanquel paused and took a deep breath. "But then she said something quite peculiar."

I felt like I was about to lose it. "What did she say?"

"She said she adored you, but she believed you would be happier if she had a more stable career, got a regular job, and considered painting more of a hobby than a profession."

"She said that?" I wasn't shouting, but I felt like I might start at any moment.

"Of course, I told her that you would be happy no matter what she decided to do, and you always speak highly of your deep admiration for her incredible talent. But from my own experience with my father, I understand how hard it can be for someone from the outside to truly grasp the life of an artist."

"Stop right there, Sanquel," I shouted, raising my hands. I took a few deep breaths to regain my composure. "I ask her all the time about her work. Whenever she shows me something she made, I tell her how fantastic it is. I've never said a discouraging word to her, ever!"

Sanquel nodded solemnly.

"But," I was still shouting, "the one, single, solitary time I told her about what I'm trying to accomplish in my job, my vocation, my life's work, my legacy, when I tried to explain my goals, she tore into me and told me I should be ashamed of myself."

I hadn't realized how deeply it hurt when she criticized my work, calling it just another way for wealthy people to get an edge over the poor. I felt her critique cut right through me.

"Peter was an irate young man when I met him," Sanquel said, interrupting my brooding thoughts. "Sometimes I think he doesn't realize I saved him from his demons. I wanted to do what was right. And that has worked out well for both of us."

Why was he talking about Peter? I was about to lose it and have a complete breakdown right then, and Sanquel was talking about saving his gay lover.

"Sophie is struggling right now, but we, you and I, can help her achieve her dreams," Sanquel said with a smile. He truly had a fantastic smile. "We can save her together."

"I've tried," I said, choking up. "I've tried to help her, but she's too stubborn to accept my help." I hadn't realized how much pain I'd bottled up. "If I treated her like she treated me, only listening to half the things she says, she would have dumped me long ago. She's lucky that I love her so much."

"But she fears that she will become dependent on you," Sanquel said. "Peter was the same way, refusing my help because of pride. Peter is not the best listener either. Trust me. But this is what I want to share with you. I have a plan for Sophie that I believe she will be willing to accept. I told her about my good friend, Heather Gonzales, who recently opened a gallery in the heart of Brooklyn. I called Heather this morning, and she said she'd be happy to speak with Sophie. Heather can get Sophie the exposure she needs to grow into a successful artist."

I wiped my eyes. I seriously doubted Sophie would make the call. She was too damn proud for her own good. "Thanks," I said. But as I headed up the stairs, a question nagged at me. Why the hell was Sanquel trying to help Sophie anyway? He should mind his own business. Sophie was my girl.

On Friday, I hurried to finish some last-minute tasks and headed to Chicago. I promised myself I wouldn't think about Sophie, but, as you might guess, I couldn't stop thinking about her during the five-hour drive up I-55 in the heavy rain. I kept wanting to call her to wish her a happy birthday, but I was too upset with her. The last thing I needed was another fight.

I kept obsessing over Sophie's visit to Sanquel. He had shared the basics of the encounter, but my mind kept trying to fill in the details. Why had she come to the house that day? She knew I would be at Ian's party. What was she trying to do? Was she attempting to seduce him? I know that sounds crazy because Sanquel was gay and already with someone else. But I could see the temptation. Maybe his 'predilection' made him seem out of reach—a challenge. Did she think she could win him over with her youth, beauty, and charm? It had worked on Ravi, so why not him?

Or was she just there for sympathy? 'Narin's nice enough but lacks understanding and appreciation for the life of an artist.' Even if she ignored me while working, she definitely liked to unload her stress and misery on me. I was mystified that she would complain about everything and then reject all my help. Her irrationality was simply unbelievable.

But I wasn't done tormenting myself. During the last hour of the drive, I thought about the paintings of Sophie in Ian's red room. I pictured her long, bare limbs, her slender waist, and her small, rounded breasts. The paintings evoked in me a powerful and painful mix of lust, revulsion, and jealousy, making it hard to stay still while I was driving. It didn't help that I could barely see the road because of the sideways rain. I was pulling at my hair so forcefully that it nearly pulled out of my scalp.

I couldn't bear the thought of Sophie, my Sophie, sitting passively while Ravi studied the intimate details of her body. She looked so young in those paintings. Had Sophie posed for Ravi when she was still his student? And why hadn't Sophie told me those paintings existed? Did she think it was irrelevant to our relationship? Who knows what other secrets she was hiding from me?

After that, my mind spiraled out of control. What if Ian sold the paintings, and they ended up in a public gallery where everyone could see them? They might be reproduced, published, or even displayed on billboards. My parents could see them! No, there was no way my wife would be fully exposed for the world to see. I had to end things with her. That was certain. I had no other choice. It was a mistake to get so close to her. The thought of ending our relationship felt like a huge weight was lifted off my shoulders.

But then I pictured her again, streaked with paint, her face twisted in concentration as she attacked the canvas. The truth is, I was completely head over heels in love with her. Sophie was an impoverished art student with nothing to sell but herself. Morey was a stingy bastard. She desperately needed money. Ravi exploited Sophie's poverty and innocence. Ravi knew the power he held over her. "I hope you suffered a horrible death," I yelled.

The wheels turned in my mind. Maybe I could persuade Ian to sell me the paintings. I'd offer him a fair price; he wouldn't refuse. After all, we're best friends. I nodded as I considered the idea. Ian would understand my concerns. But then, a new thought chilled me to the bone. What if more of these paintings of Sophie were hidden in private collections or a warehouse somewhere? Maybe there were even more vulgar and pornographic paintings than the three Ian had.

I thought I was going to be sick. How could we have a happy life together when a sudden revelation might throw everything into chaos? *I needed to stop pulling at my hair.* A raw, tender patch was forming on the back of my scalp. I knew I would have to end things with Sophie soon. When I returned to St. Louis on Sunday, I'd tell her our relationship was over. I was sure she'd also feel relieved, and my scalp would thank me for it.

I took a few deep breaths, and the rain seemed to let up. The issue was settled.

My mind raced through the events of the past few weeks: Ian's friendship, Maru's hostility, a chance to restart my career with my new job, the ethical dilemma of Operation Juggernaut, the looming political upheaval at Harvester, and my doomed relationship with Sophie. Every time my chest tightened, I took a deep breath and exhaled to ease the tension.

I thought about the Desert Prince and sparring with Monkey. Men need to do things like that: challenge their muscles and wits with high-stakes battles. We feel strong when we solve our problems by squaring off in mortal combat. Half the world's problems could be fixed if men did that regularly. I missed the sparring badly and hoped the Desert Prince would visit me soon.

Thirty-seven

Due to the rain, I arrived later than I had expected. As I approached my parents' front porch, I saw a fly caught in a spider web, glistening with raindrops that clung to it. I used the edge of my key to free the fly, but its body dropped to the ground. The fly was already dead, just an empty shell. I knelt down to look at it, but a gust of wind swept it away.

I rang the doorbell, and Mom answered. She hugged me. "You look good. I can see you've been eating well," she said after a stern appraisal.

"I've been exercising too, lifting weights," I said, flexing my biceps.

Deepa stood behind Mom, and I hugged her. "I bet I know why," Deepa whispered in my ear.

You have no idea.

Behind Deepa stood a tall, gangly Indian guy who looked about seven or eight years younger than me. He wasn't bad-looking—some Indians appear as if they just got off the boat—but he seemed more "American" with his hipster glasses, dress shirt, tie, and slacks.

"You must be Rajeev," I said.

He shook my hand warmly. "I've been looking forward to meeting you. Your sister had told me all about you," he said.

"Don't believe any of it. I'm really not that bad."

Rajeev laughed nervously, and his enthusiasm seemed forced, but I appreciated his effort.

"Come and sit down in the living room," Mom said. "Chat with Rajeev and Deepa."

"Where's Baba?" I asked.

"He's still at work. He's been working too hard lately. He'll be home soon."

Mom slipped away to the kitchen.

"I brought some presents," Rajeev said. "Your mother suggested I put them in your room, Narin. I hope you don't mind."

"Sounds great." I struggled to focus on Rajeev's words. After thinking about Sophie for five hours, my mind was exhausted. We went and sat in the living room.

"I was admiring the icons in your room," Rajeev said. "They're stunning."

"I was showing Rajeev the wooden statues that our grandmother brought from India," Deepa explained. Her eyes shifted from Rajeev to me. "Narin," Deepa said sharply when she noticed I wasn't paying attention. "Rajeev knows everything about Indian mythology. He spent two years in India studying at an ashram before starting medical school. Isn't that interesting?" Her eyes pleaded with me to act like a normal person. I tried to find the strength to do so.

"Now we must be careful, Deepa," Rajeev interjected. "Mythology is not the right term for our beliefs. Our faith should properly be called a religion."

I shot a glance at Deepa, raised my eyebrows, and smiled. I shouldn't have done that; it wasn't very nice. Her face was flushed, and I wasn't sure if it was from embarrassment after being corrected by her boyfriend or from resentment toward her horribly insensitive big brother.

For some reason, I felt freezing cold. I slid my hands under my thighs. "That's very interesting," I said. "Tell me more."

Rajeev smiled and adopted a different tone, one that resembled that of a teacher, a prophet, or perhaps a saint. "We have an unbroken tradition that dates back thousands of years," Rajeev said. "We must honor and respect that tradition. It is a tradition of learning and scholarship. That's why we see a divine order where others see chaos. But our generation must ask itself: can we live up to the great examples set by our fathers and their fathers? Can we live a life imbued with the divine? Surrounded by modern temptations, can we live a life without worldly attachments?"

"Such lofty questions," I replied, smiling at my sister.

The more she scowled, the more I smiled. Old habits are hard to break.

Rajeev nodded thoughtfully. I considered provoking Deepa further by asking Rajeev about his religious beliefs.

I haven't brought this up before, so now might be a good time. Our family wasn't religious. Some people strongly dislike religion and its ideas, but that wasn't us. We simply never discussed it. We were nominally Hindu, but that didn't influence how we lived.

Our parents sometimes took us to the Hindu temple for special celebrations or holidays. The temple we visited was impressive and built to resemble one in India. I'm sure it was designed to inspire awe among the faithful. However, my parents went more for social reasons than for religious ones.

Deepa and I used to run around during the ceremony with the other kids when we were little. We didn't know any better. As we grew into sullen teenagers, we sat with scowls and our arms crossed, waiting for it to be over. I was surprised to learn that my grandmother, my dad's mom, was hyper-religious. Maybe that was normal if you grew up in India.

The truth is that I knew much more about Christianity than Hinduism while I was growing up. I learned everything about Jesus from my evangelical classmates, who saw me as friendless and pitiful—an easy target for conversion. I never felt they genuinely cared about me or my immortal soul; it seemed more about hoarding riches in heaven for themselves. Still, I never took them seriously.

My dad came home, and we headed into the dining room for dinner. Dad sat at the head of the table.

"It is so wonderful to have you all here," Mom exclaimed. "Narin, Deepa, and now Rajeev. What a wonderful homecoming."

"Your father is a physician in Los Angeles?" Dad asked.

God. Please. Don't start with the interrogation already.

"He's a cardiologist," Rajeev replied.

"Is that what you wish to be?" Mom asked.

I was starting to feel uneasy as I sat through a son-in-law interview I couldn't care less about.

"It's still a long way off," Rajeev replied, "but I hope to become an oncologist someday."

"That is an excellent field," Mom said. "You know that two of Deepa's grandparents died of cancer."

"It is a devastating disease," Dad added.

"I believe it's a field where I can help many people," Rajeev said, "and from what I understand, it's well compensated for a medical specialty, though I don't consider that very important."

"That is certainly one factor to consider," Dad said. "It is important to make sure that you are comfortable."

"I would like to pursue medical research," Rajeev said. "There are so many fascinating innovations in cancer treatment."

"I'm sure you'll succeed in whatever field you choose to pursue," Dad said.

I wasn't sure why he was so confident about this, but I didn't ask. I twirled my fork in my rice, completely lacking any appetite. The only reason I wasn't having a full-blown anxiety attack was that I'd decided to break things off with Sophie, no matter how painful it would be. That meant I didn't have to tell my parents about her. They're not the type to blow up and yell; that would have been easier. Instead, they would give me disappointed looks. I'd already let them down once by choosing not to go to medical school. I hated the thought of disappointing them again.

"Do you enjoy living in St. Louis?" Rajeev asked.

I lifted my gaze from my plate. They were all staring at me. "I do," I replied automatically. "It's affordable, and the traffic isn't too bad."

"Are you planning to stay there?"

My cellphone rang, and Sophie's photo appeared on the screen.

"I have to take this." I pushed away from the table. "It's company business."

Sophie tried to speak, but her words turned into sobs as she gasped uncontrollably.

"What's wrong? Why are you crying?"

Sophie tried once more. "Dad's dead."

"Morey's dead?" I stepped back from the dining room, not wanting my parents to overhear.

Sophie replied, "One of his neighbors called the police because they hadn't seen him for a few days."

"How did he die?" I whispered.

"They say that he died—"

"Sophie?"

"They say he died from the switch," she sobbed. "Was he taking Red Sky?

"I—I wasn't aware of that."

She cleared her throat. "You know, Narin, Dad had these super sharp kitchen knives." She started sobbing uncontrollably. "I don't understand. How could he have died this way?"

I pictured her with her eyes shut tightly and her body trembling.

"It's my fault, Narin. It's completely my fault. I've been so cruel to him."

"No," I said. "It's not your fault."

"After Mom died, I should've listened to him. I should've been more kind. I should never have cut him off. I should've begged him for forgiveness for hurting him so badly. I could've been a much better daughter. I—I'm such a horrible person. I can't believe it."

"You are not a horrible person," I said firmly.

"I just don't know what I'm going to do," she said. "What am I supposed to do? Like the funeral and everything?"

Mom called from the other room, "Is everything okay?"

"I need to go, Sophie," I said. "I'll call you back later."

I called Ian, but he didn't answer, and the call went to voicemail.

I went back to dinner but didn't eat anything. I had no idea what to do. I needed to get back to St. Louis as quickly as possible.

It was crystal clear to me what had happened. Morey was a drunk, but he wasn't a Red Skyer. Someone had injected him with a dose of Red Sky and H138. The police often see people who hack themselves to death from the switch. As gruesome as it was, such scenes were common during the Red Sky era. They would test him for DMTA, but they couldn't possibly know about H138. What a sick, sadistic, and brilliant way to murder someone.

Then I remembered what Ian had told his dad. "Don't worry about Morey. I'll take care of him." Now that Operation Juggernaut was over, Morey had to go. I struggled to picture Ian doing this. Ian was a schemer, but I couldn't see him overpowering Morey and injecting him. Yet, that seemed like the only option that made any sense. And if Morey was expendable, then what did that make me? *Shit. Shit. Shit. What the hell was I going to do?*

I was shivering as I called Sophie again to comfort her and gather more information.

Her voice carried a weary, shell-shocked detachment. "I've been getting all these phone calls from Australia. I guess bad news travels fast," she said with a small, wet laugh. She sounded drunk, her voice gurgling and her speech slurred. She told me what happened. The police went to check on Morey. They had to break down the door when he didn't answer. The officer didn't want to go into the details of what they found but told Sophie that 'all indications suggest that he died of the effects of DMTA-induced psychosis.'

I promised to return to St. Louis as soon as I could. She acknowledged this with a simple and distant "Okay." I was worried she might hurt herself that night. My first instinct was to go back to St. Louis immediately so I could be there to comfort, grieve with, and protect her. But then I thought, somewhat vindictively, that she had rejected my help before. Maybe now she would realize she needed me. I decided to wait until morning.

I was lying in bed when I had a realization. The Desert Prince had been on my grandmother's shrine. I hadn't looked at the figures closely when we put them away, but it was all coming back to me. I'd held him, or at least a wooden likeness of him, the Warrior Prince, with a sword held high. Somehow, this struck me as amusing: me manhandling the Desert Prince.

I opened my sock drawer, and there he was on top. I pulled out the wooden statue and examined its finely carved details. Later that evening, after my parents went to bed, I went downstairs to my dad's study. I turned on the lights and looked around the room. There he was, Monkey, exactly where I had seen him before. I took the Desert Prince out of my pocket and placed him next to Monkey.

I stood there, looking at them and thought about all the hours they had spent training me. They believed I was worth their time and effort. A tear rolled down my face. I folded my hands and bowed my head. I wanted to say something, maybe a tiny prayer, but I couldn't think of any. "Thanks," I whispered and wiped the tear away.

I saw my sister trimming her nails on the couch in the living room, so I sat down next to her. She was wearing an oversized white T-shirt and plaid shorts, with her hair covering her face. She lifted her feet to trim her toenails.

"You didn't talk much tonight," she remarked.

"Sorry," I said.

"I think Rajeev was intimidated by you. He seemed nervous," Deepa noted.

"Sorry if I seemed intimidating," I said coldly.

"What was that phone call about?"

"Some shit's going down at work. I'm trying to find out what happened, but I haven't had any luck. I'll probably need to head back early tomorrow."

"Are you working with that sociopath who used to be Dad's student?"

"What do you know about it?" I hissed.

"Mom told me all about it," she said. "Mom says that Dad should never have let you work with him. She said that he's a terrible human being."

"I can do whatever I want," I said. "It's my life."

"I'm just telling you what she said."

I stared at the mirror hanging on the wall across the room. It felt like an eternity since I first saw Ian and Maru in that mirror. From where I sat, I couldn't see my reflection.

"Do you remember when our cat killed your hamsters?" I asked spitefully. "Remember how she ripped open the chest of one of them and how she bit the other one's ear right off and tore off half his face?"

She didn't say anything. She just kept trimming her nails.

"That's what life is like, Deepa," I shouted. "That's what the real world is like. It's not about living a life imbued with the divine."

She looked at me with fiery eyes. "Fuck off," she yelled. "You are such a total dick." She stormed out of the room.

The house was quiet. I went to my bedroom, turned on the TV, muted it, flipped through hundreds of channels, and then turned it off again. I started pacing. Since childhood, I've known every spot where the floor creaked. I jumped from one spot to another as if playing a tune. I opened the window, sat on the floor, and looked out at the twisted branches of the maple.

Deepa would have to defer to Rajeev's opinion on all the important issues—God, religion, and raising children—that much was clear. She wouldn't have room to develop her own beliefs. He would treat her well and be a faithful and loving husband, but she would feel smothered, stifled, and belittled. She is too young to see the big picture. I got up from bed, determined to explain all of this to her. But then I stopped at the first creak of the floorboards. Who am I to give anyone relationship advice? I returned to bed and stared at the glow-in-the-dark stars on the ceiling until I fell asleep.

I stood on the edge of the cliff, looking out over the red desert. A full moon hung in the sky. Monkey wasn't there, only the Desert Prince. He, too, was gazing at the endless dunes under the bright moonlight.

"I'm in trouble," I said. "I don't know what to do."

"There is no problem so great that cannot be solved with the edge of a sharpened blade," he replied without looking at me.

I knew he would say that.

"Are you ready?" he asked.

"For what?"

He drew his sword. I stepped back, pulled the knife from my pocket, and prepared for battle.

Thirty-Eight

He was as tall as Monkey and even more intimidating. He wore a red robe, a brown tunic, and a golden chain with his amulet. I thought his necklace looked like a vulnerability as I sized him up. I could grab it and choke him. But, damn, he had a sword.

You might think I'd be terrified, but I was exhilarated. I was finally doing what I'd trained for. He examined me, and if he saw any weakness, he didn't show it. I moved back to a safe distance as he approached, waiting and watching. He rushed forward and swung his sword. I ducked and lunged to my left. I noticed another weakness: his robe was bulky. If I could grab it, I could maneuver onto his back and slit his throat. He repositioned himself. He was an experienced warrior, probably centuries old or more. He wasn't going to try his hardest; that would be unfair. But I wasn't counting on his mercy; I was determined to give it everything I had.

He raised his sword and charged again. I sprang upward and backward, reaching for the folds of his robe but just missed. We fought for a while—him attacking, me dodging. I hate to brag, but I was holding my own. I felt strong, agile, and capable. Most importantly, I had a plan.

He lunged again, and this time I grabbed his robe and clung on for dear life. He thrashed, trying to shake me off, but I had a firm grip and wasn't letting go. He aimed his sword at me, but I was now behind him, on his back. I seized the gold chain and pulled. Though this didn't bring him down, it managed to throw him off balance. I quickly moved to his front and slashed at the arm holding the sword. A direct hit! Snake defanged! He looked shocked by this. I leaned in, gripping his collar, our foreheads touching, his breath hot on my face.

"What is your name?" I demanded.

He smiled and said, "Finish what you have started."

I thrust the knife into his chest.

He didn't bleed or fall; he just stood there, as pleased as punch. "You are ready," he said.

I stepped back. I don't know why I did this, but I knelt on the ground before him and bowed my head. Tears streamed down my cheeks, not from sadness but from overwhelming joy. I should have asked him what I was ready for. I know I was foolish not to ask. I don't know if he would have told me. Somehow, he knew I would face a life-threatening challenge where I'd have to use the knife to kill my enemy. When we first met, he told me that I was the last in the lineage of his beloved teacher. He needed me around to keep the line going. But who was going to try to kill me? And why?

All of this brings us to Saturday, July 8. We're getting close to the end now.

That morning, I woke up early and explained to my exasperated parents that I had to return to St. Louis immediately due to a work emergency. I waited for Deepa to come downstairs and say goodbye, but she never did. As I pulled out of the driveway, I saw my mom standing on the porch, frowning.

While driving back, I tried to reach Ian by phone, but each call went straight to voicemail. I desperately needed to find out if he was responsible for Morey's death. But what if he was? I tried to imagine how I would handle that. What if Ian told me he had 'taken care of Morey'? Would I report it to the police? Would I tell Sophie? I realized I was in a serious bind. I'd hitched my wagon to Ian, and my fortunes depended on him. If he went down, so would I. I might even be considered an accomplice.

But what if I didn't report the crime? Then I really would be an accomplice. I thought about calling Sophie but decided against it. I didn't want to deal with a blubbering drunk on the phone. I'd see her soon enough. This was better handled in person.

When I returned to St. Louis, I went home first, but Sanquel wasn't there. I quickly showered, changed into clean clothes, and gathered my thoughts. Then I headed to Sophie's house. I was nervous about what I might find. I expected her to be a drunk, emotional wreck, but, to my surprise, she greeted me with reddened yet dry eyes and a warm hug.

"Thanks for rushing back," she said. "I'm so glad to see you. The funeral will be on Tuesday. I've been on the phone with Ian all morning."

"With Ian?" I wasn't sure I heard her right.

"He's been incredibly helpful." She released me. "He handled all the funeral arrangements, paid for everything—I mean, I'll pay him back as soon as I can—and he even got a lawyer to help me settle Dad's estate. Dad didn't have a will or anything. His house was in foreclosure, so Ian says it now belongs to the bank. Ian wants me to go through Morey's belongings and keep anything sentimental, like small keepsakes. The rest will be auctioned off to settle his creditors. Ian said he'll handle all the arrangements. Another important thing Ian told me is that Dad's creditors can't pursue me for Morey's debts. I was so relieved to hear that. Ian has been a blessing. I don't know what I would've done without him."

My hackles were up. If only she knew the truth. Ian was almost certainly the one responsible for Morey's murder. He wasn't her savior. But I had no proof, so I kept my suspicions to myself.

That afternoon, we drove into Illinois on Highway 40 to Morey's house. The police had finished collecting evidence and said it was okay for her to go inside.

"I called an art dealer friend of Sanquel's last week," Sophie said casually as I drove. She didn't mention that she'd visited Sanquel while I was at Ian's party.

"It was a complete disaster," she said.

"Really?"

She continued, "Some snooty-sounding guy answered the phone. He acted like he was doing me a favor just by picking up. I told him my name and asked to speak with Ms. Gonzales."

"Who's that?" I asked. Of course, I knew, but I wanted her to think I didn't.

"She owns an art gallery in Brooklyn," she said. "Anyway, I was on hold forever. I was about to hang up when she finally picked up the phone. I could hear her long nails clicking on the keyboard as we talked. I don't think she was even really listening to me."

"Wow," I exclaimed. "That's so rude."

"I told her Sanquel recommended I call her," she said. "I explained that I am an artist from St. Louis and want to exhibit my work at her gallery."

"What did she say to that?"

"She laughed and said, 'But of course you do.' I felt like a complete idiot. She said she had spoken with Sanquel about me. Then she asked me how much inventory I move."

"What does that even mean?" I asked.

"I didn't know," she said. "When I didn't reply, she asked, 'How much do you sell, let's say, in a month?' I told her about $1,200. That's the most I've ever made, but still, I thought that was fair."

I nodded.

"Then she told me that she has artists who make ten times more than I do, beating down her door to get their work displayed in her gallery. I was getting angry now. So, I said, 'I thought a gallery's purpose was to find promising artists and give them a venue to reach a new audience.' And do you know what she said to that?"

"What?"

"She said that the gallery's job is to make money. She asked me to send her a copy of my portfolio for review and said that the New York market is much more competitive than a small town like St. Louis. She doubted anyone in New York would be interested in my work. And get this, then she said it would be unfair to other artists, to the gallery, and even to me to put me in a position where I was unlikely to succeed."

"What a bitch!" I said. What this Gonzales lady said to Sophie was almost word-for-word what McGregor had said to me not too long ago. "Well, fuck her then," I said. "It's their loss. Your work is amazing."

"I should've told her off," she said. I could see the gears turning in her mind; if Ravi were still alive, he'd have used his influence to persuade Ms. Gonzales to display her art. But Ravi was dead, and she had no one to help her.

We drove through small towns and saw endless fields of soybeans, corn, and other crops. Finally, we left the highway and turned onto a paved but poorly maintained road. After several more miles, Sophie instructed me to turn down a long, unpaved driveway bordered by overgrown shrubbery on both sides. I wasn't happy about having to drive Ian's car on such a rough road, but what was I supposed to do? The drive ended at a small, white house that looked lonely and neglected. I felt somewhat anxious about what we might find inside.

"This is where I grew up," she said in a dazed voice. "It's been years since I was here."

Inside, there were no obvious signs of chaos. The living room had worn, mismatched furniture, but it felt cozy and inviting. I caught a faint smell of bleach and wondered if the cleaners had already been through. Sophie picked up a photo from the mantel that showed a dark-haired toddler flanked by a young Morey and his stunning wife. Morey had dark, unruly hair that extended into long sideburns. He wore a double-breasted suit with padded shoulders and wide lapels. Sophie's mom wore a short plaid dress. She looked so much like Sophie that I thought it could be a picture of her. The only difference was that her mom had a permed head of voluminous hair. Sophie tossed the photo into a cardboard box and went upstairs. At the top of the stairs, Morey's bedroom was marked with police tape. I thought about looking inside but ultimately decided not to.

Sophie went directly into what had to be her childhood bedroom. She rummaged through the closet and threw stuffed animals and toys onto the twin-sized bed. The air was stale, so I opened a window. A warm breeze came in, carrying the scent of hay, dung, and pesticides. I sat by the window and looked out at the gray silos, red barns, and white farmhouses scattered across the landscape.

You might wonder what happened to my plan to break up with Sophie as soon as I returned to St. Louis. Well, Morey's untimely death threw a wrench into those plans. I couldn't break up with her right after her father died. I'm not a callous jerk, after all. I'd wait a week and then tell her. That was the plan. Until then, I'd pretend nothing was wrong.

"There was a group of us girls in junior high," she laughed. "We'd walk past all the cute boys' houses and moon them. We'd pull down our pants and wave our butts around. We didn't even care if they saw us or not. It was thrilling all the same."

"Sounds like fun," I said. *How could I marry a girl who does something like that?*

I'm not sure why that small piece of information upset me so much. I guess it brought back memories of how lonely I felt as a kid. I was one of those losers that all the tough kids would pick on to prove how tough they were. I'd get hit, pushed into lockers, and spat on. It was a living hell. If only I'd known that none of these people mattered. If only I'd realized they were the losers, not me. I was so close to being perfect. Once the three-drug cocktail was available, I knew I would be unstoppable. We finished our work and headed back to St. Louis.

On that Tuesday, July 11th, Harvester shut down so everyone could attend Morey's funeral. Sanjay Chandra delivered a heartfelt eulogy, his voice trembling as he recounted their long friendship. Several other eulogies followed, sharing stories about the company's founding, the numerous trips to South America, the many misunderstandings with native peoples and government officials, the near disasters, and the remarkable successes.

I looked around and saw Ian hanging out at the back. I excused myself from Sophie and went over to Ian, who gestured toward a small nook in the lobby.

Once we were out of sight, I asked Ian sharply, "What the hell happened?"

"How would I know?" he replied. "I'm just as shocked as everyone else."

I moved closer and pressed against Ian's chest. "He wasn't a Red Skyer, Ian. Someone killed him."

Ian's eyes darted as he looked around the lobby behind me.

I opened my mouth to speak again, but Ian grabbed my shoulder. "No more of that, Narin. Be careful not to bite the hand that feeds you. That aphorism applies to everyone except my bird. Do you understand?"

I looked at the hand holding me. "I thought we were best friends, Ian. And you're threatening me again?"

Ian let me go. "I didn't kill him," he whispered. "And I can't believe you'd think I did." He took a deep breath. "I liked Morey and considered him a friend. I'm just as upset as everyone else that this happened." He stepped closer until he was breathing in my face. "But if you look at this rationally, you'd see that the timing is good for you. Morey was an obstacle to you. You told me so yourself."

"Bullshit," I spat. "Don't act like I wanted this. I'm in no way responsible."

"It's not about wanting anything," Ian replied. "It's about adapting to the new reality. You may not want to face the truth, but Morey's death works well for your career and your relationship with Sophie."

"I refuse to accept that." I was about to tell Ian that Sophie and I were done, our relationship was over, but I decided not to say anything for now. Besides, it was none of his business.

Ian sighed. "Listen, Narin, you need to calm down. Everything's going to be okay. There's a stockholders' meeting next Tuesday. Things are going to change at Harvester. I'll tell you all about it on Monday. Regardless of the issues, we'll get through this, you and me. Failure is not an option."

"Morey told me he was getting a hundred-thousand-dollar bonus because our project was successful," I said. "That money should go to Sophie."

Ian nodded slightly, then smiled warmly and looked behind me. I turned my head and saw Sophie approaching. She passed me and went straight into Ian's arms.

"Oh, Ian!" she exclaimed as she hugged him. "I can't thank you enough for all your help. I would never have managed without you. How can I possibly pay you back for your kindness?"

Ian kept eye contact with me over Sophie's shoulder until she finally let him go.

Maru watched all of this from across the room.

My blood boiled with jealous rage. It's terrible not to be needed by the people you love.

I decided to always keep something in hand that people desperately needed. You're an obstacle if you don't bring something to the table, and obstacles are meant to be removed. I would make myself indispensable, impossible to ignore. When I got home, I grabbed the knife and went down to the basement. I spent an hour lifting weights, then practiced with the knife. I pictured my enemy standing in front of me, dodging and striking with deadly force. By the time I finished, my clothes were soaked with sweat.

Thirty-Nine

The next day, Ian called me to ask why I hadn't stopped by for coffee with him. I explained that Morey's death had thrown off my routine. I didn't mention my fear that he might poison me to have Sophie all to himself. He insisted I come to his office anyway, and I couldn't refuse.

When I arrived, Ian had already made two cups of coffee. I sat down, grabbed my cup, but didn't drink it. He sat down beside me.

Ian said, "I have a water stain on the ceiling above my bed."

Where was this heading?

"It started small, as a barely noticeable reddish dot," he said. "But from that innocent spot, it expanded and spread out in all directions. I was looking at it last night. It had taken on a greenish tint in the dim light, with hints of brown." Ian took a sip of coffee and continued, "It reminded me of a field next to my childhood home where I used to collect bugs in a jar. One year, I planted sunflower seeds and watched them bloom all summer." Ian smiled and took another sip. "It was my favorite place as a kid. An occasional stray dog or cat would appear, and I would pretend they were mine. My parents didn't allow me to have pets."

I didn't say anything; I simply watched him sip his coffee.

"I wondered how long it would take the creeping stain to spread across the ceiling and cause it to collapse on me," he said. "Maybe then I could return to that field." He looked up, startled, as if he'd forgotten I was there. "But I couldn't go back there. It was parceled off long ago and made into a subdivision of identical houses."

The coffee cup was hot, so I set it down. "Is there something you wanted to talk to me about?"

He smiled. "There are going to be big changes soon," he said.

"Like what?"

"Well, for starters, I've arranged for you to be the interim director of the Neuroscience Division."

He watched me for a reaction. I was stunned, and I'm sure my face showed it.

"Wouldn't someone with more experience be a better choice?" I stammered.

He shook his head. "We want you, Narin. We will announce this at your division meeting later today."

Oh shit.

By the time the meeting came around, I had worked myself into a panic. I was sweating and feeling faint. Several spots on my scalp ached from all the hair pulling. I knew the news that I was the new interim director of the Neuroscience Division would land like a lead balloon. And rightfully so. What was Ian thinking? Could I refuse the position? What would Ian do if I turned it down? Would I then become dispensable? Then I remembered that being dispensable was no longer an option for me.

People were talking about Morey and sharing stories as anticipation filled the room. Everyone was waiting for an announcement. I started tugging at my hair again but stopped when I saw Ian walk down the hall.

He paused at the door. He should have entered the room, sat down, and taken questions, but he didn't. Standing in the doorway, he remained silent. All eyes were on him. The room grew quiet. He cleared his throat. "After much deliberation," Ian said, "the Executive Committee has decided to name Narin Roy the Interim Head of the Neuroscience Division."

You could have heard a pin drop.

Ian continued, “I know this may seem like a surprising choice, but we plan to focus our drug development in this division on treatment options for the Red Sky epidemic, and we believe that Narin is best qualified to lead this effort going forward.”

“Is this Chandra’s decision?” someone shouted. I couldn’t tell who said it.

“The Executive Committee has made its decision,” Ian replied calmly. “And that decision is final.”

Ian left, and the room erupted into urgent conversation. I stood on unsteady legs, raising my voice to be heard over the crowd while trying to maintain some semblance of dignity.

“Um—I am as surprised as you are by this decision. I didn’t expect this.” I cleared my throat. “It is—uh—a great honor to be chosen for this position. As the new leader of the Neuroscience Division, I will do my best to serve all of you and foster the most collaborative work environment possible. Thank you in advance for your support.”

With that, the meeting ended, and everyone hurried off in different directions. JoAnn and Ahmed didn’t even glance my way. No one stopped to congratulate me or even insult me. They regarded me as not even worth their breath. I was left alone, a leader without followers.

I went home early because I had a bad headache. I thought about calling Sophie to tell her about my surprise promotion, but it felt like too much to handle. I would take a few migraine pills and call it a night.

When I got home, I saw Sanquel sitting on the couch working on his laptop. I hoped he would ignore me, but he motioned toward the sofa and asked me to sit with him, so I did.

“You look worn out,” Sanquel observed. “Work taking its toll, my friend? Did you enjoy your trip to Chicago?”

“You would never believe what has happened these past few days. I can hardly believe it myself,” I groaned.

“Please enlighten me,” Sanquel said, closing his laptop and turning his attention to me.

"Well, for starters, Sophie's father died unexpectedly," I explained. I had no intention of revealing the gory details.

Sanquel's eyes widened, and he exclaimed, "How dreadful for her! How is Sophie taking it?"

"About as well as could be expected," I said. "The funeral was yesterday. Most of Harvester attended."

"Oh, I almost forgot. He was your boss at work, right? What was his name again?"

"Morey Whitely," I replied. "He was the head of the Neuroscience Division at Harvester. His death has already caused a lot of changes."

"That must be very unsettling, especially since you are new to the company. Do you know who will replace him?" asked Sanquel.

I hesitated before deciding to tell him. "As I said earlier, you would never believe it. Today, they named me the interim head of the Neuroscience Division."

Sanquel's jaw dropped as he stared at me for several long seconds. "Please explain this to me! How did this happen? I guess congratulations are in order."

"I'm not so sure about that," I replied. "This promotion has made me the least popular kid in the schoolyard. I'm sure I have a target on my back." I looked over my shoulder for emphasis.

Sanquel shook his head as if trying to clear his confusion about this turn of events.

I continued, "I'm sure it was a political move to signal changes in company leadership. The appointment wasn't made because of my stellar leadership skills."

"Does Sophie know?" Sanquel asked with concern.

"About the promotion? No, not yet. I'm not sure how she'll react to me taking her dad's old job the day after his funeral."

"I do not envy you, Narin. Can I help in any way?" He grasped my hand warmly, but I reflexively pulled away.

"I'm not sure what you could do. I'm just in over my head." Sanquel had no idea how deeply I was in over my head. What I had shared was just the tip of the iceberg. I went upstairs, took a shower, and crashed into my bed.

I spent the rest of the workweek trying to settle into my new job. Most of Thursday was spent visiting each of the six neuroscience labs. I initially checked the roster, trying to memorize all the names. I'm terrible with names and eventually gave up on that. Instead, I decided to just wing it.

I entered each lab with a big smile and a firm handshake. My plan was to say some cliché things about moving forward and teamwork. I was met with indifference at best and open hostility at worst. The hostility was most obvious from JoAnn and Ahmed, who turned their backs on me as I entered. They didn't say a word when I spoke to them. I knew they were angry, and I tried to make amends, but they wouldn't listen. So I gathered my belongings from my third-floor room and took them to the Executive Suite.

On Friday, July 14th, I went through Morey's inbox and tried to understand my new role. I had no idea what problems Morey was facing. I spent the day signing forms and approving expenses. He didn't have an assistant and kept no records, only scattered phone numbers scribbled on Post-it notes. I spent the entire day figuring out where to fax forms and who to contact, and I made a mental note to ask Ian how to get a secretary to help me. *No poker game for me. I was exhausted.*

I was just about to call it a day when I heard a knock at my door. I figured it was probably Ian, so I shouted, "Come on in." To my surprise, it was Maru. He stepped inside and closed the door behind him.

"What can I do for you, Maru?" I asked. Honestly, having him in my office made me nervous, especially with the door shut. "I was just about to leave," I said as firmly as I could.

"It's time for us to talk, Narin," Maru said, sitting across from me. "Spare me a few moments."

I chose to listen to him because, as the VP of Drug Development, he was my direct supervisor.

"I know you and Ian have become quite close," Maru began. "That's why I need to talk with you. You need to know the truth so you can make the right decision."

I leaned back in my chair. "Go ahead, educate me," I said.

A hateful smirk spread across his face. "I am aware that you and Ian have been plotting to take control of Harvester."

I started protesting, but Maru interrupted: "Shh, shh, shh. Calm down, Narin. I'm here to set things straight. Ian has offered you a quick path to the top of Harvester's hierarchy. I'm sure he promised you fame, fortune, friendship, and love. But I'm here to give you a different way. I'll show you the right way forward."

I wanted to respond, but I couldn't come up with anything to say.

Maru continued, "We're brothers, you and me. We are more alike than you realize. Our fathers have both struggled and suffered in this country, and for what? To give us a better life. You realize that my father spent twenty years building this company. He worked every day and on weekends. I never saw him between his trips to South America and his time here. My mother never saw him either. That's because we weren't his family. His real family was at Harvester. Morey was his real brother. Harvester's employees were his real children. By joining Harvester, you became his child. So, I must ask you, why would you ally yourself with a spoiled rich white kid who never struggled for a day in his life? Ian does not give a shit about you. He doesn't give a shit about anyone but himself."

I wasn't sure what to say because, honestly, he was right.

He continued, "I know you have serious reservations about Operation Juggernaut. Your concerns are valid. There's a fine line between good and evil, heaven and hell, kindness and cruelty. It can be hard to tell which side of that line you're on. But I believe deep down in your heart, you know which is the right side. Operation Juggernaut was a huge mistake. We should've stuck with our core business, making products that actually help people. Ian and his father are on their way out. Don't be on the losing team. You don't have much time."

I was still struggling to think of something to say, maybe making a joke about what he said or telling him he was wrong about Ian. But I couldn't bring myself to do it.

"You have until five p.m. on Monday to decide. If I don't hear from you, I'll assume you have chosen to stick with Ian." With that, Maru stood up and left.

If Maru knew about Team Blair's scheme to take over the company, Sanjay Chandra probably knew as well. Maybe he's right; I should switch sides and join Team Chandra. The truth is, I had no idea what to do at that moment. Could Sanjay Chandra really withstand the wrath of the board of directors? Would they keep him as Harvester's CEO?

I should have chased after Maru and told him I was on his side. Ian was wrong to want to steal Harvester from the Chandras. Operation Juggernaut was evil, just as Sanquel had said. But I didn't do that. I didn't switch sides, at least not yet.

If I were a good Team Blair player, I should have told Ian that Maru and Sanjay Chandra knew about their plot. But I didn't. Once I knew which side would win, I would make my move. I had to keep all my options open. I needed to end up on the winning side. Failure was not an option.

Just three days ago, on Sunday, July 16th, Sophie and I had our last dinner with Professor Goldblum. Sophie's grief seemed to weigh heavily on the old man, causing him to lose his usual eloquence. He spoke incoherently and appeared lost in thought. When we left, the professor stopped me at the door. I was worried he would bring up the truth serum again, but he didn't. That night, there was light rain, and the air was cool and crisp. Sophie carefully went down the stairs, gripping the railing with both hands. The professor and I stood in the doorway watching her.

"She's very sad," he said.

"Her father just passed away."

"Sustained happiness is impossible in this world, but she is too young to understand." His voice grew hoarse. "Happiness is fleeting, like raindrops falling on a silent lake. That's all this life has to offer. This is all we can ever hope for. She will understand this someday." He gazed at her silently as she stood on the sidewalk with her back to us. "She is too young to understand."

Forty

On Monday, July 17th, I searched everywhere for my cell phone but couldn't find it. I knew I had taken it to work and left it on my desk that morning. When I returned from a meeting on the third floor, it was missing. I'd arrived late, so I didn't have time to look for it. I spent the whole day catching up.

Maru's words kept echoing in my mind. Even though I strongly disliked Maru, I spent the day debating whether I should go to his office and tell him I was on his side. But how could I escape Ian's grip without risking everything I've worked for? And why should I trust Maru more than I trust Ian? I was so busy that by the time I finished, it was already 5:30 pm. Deadline missed.

At six, I went to Ian's office. He said he wanted to prepare me for the events at the shareholder meeting the next day. I already had a pretty clear idea of what would happen. It was a coup: Chandra would be fired, and Harold Blair would take over. That is, if their coup was successful. The Chandras knew about it, and there was a good chance they would find a way to stop it. I needed to adapt to the changing circumstances to survive this mess.

"People are unhappy about the decision to put me in charge of Neuroscience," I told him. "They won't say it openly, but they all hate me now, even Ahmed and JoAnn."

Ian chuckled. "One thing you learn quickly when you're in charge is that the pigs are always looking for a reason to grunt. Don't pay any attention. Morey never did."

"I guess I don't have a choice," I said. "There's no turning back now."

"That's the spirit." Ian slapped me on the back. "You know, Narin, I realize that my bird and I have a lot in common. We both like to rule our realm, and we both enjoy working in pairs. They hunt in pairs, you know. That's just like the two of us. It's us against our prey. That's the whole game, and it will all be over tomorrow."

"What exactly is going to happen?"

Ian was giddy with excitement. He pulled me into a seat and sat beside me, draping his arm around my shoulder. I tried to pull away, but couldn't. He leaned in conspiratorially and spoke in a hushed tone. "Dad has orchestrated this perfectly," he said. "There will be a vote tomorrow at the shareholders' meeting."

"What kind of vote?" I asked.

"The Board is demanding immediate changes at Harvester, and Sanjay Chandra is on the chopping block."

I tried to look shocked. "Isn't there anything Chandra can do to stop this?"

"Too late for that," Ian said. "He knows this is coming, but he can't do anything to stop it. We need to work things out with Maru. Morey thought that maybe if—"

A loud knock shook the door. I jumped at the sudden sound.

Ian stood up and opened the door. To our surprise, Sophie was in the hallway, gasping for breath. She rushed inside and wrapped her arms around me.

"Oh my God, Narin! You're okay!" she said frantically. "Why aren't you answering your phone?"

"I lost it this morning," I explained. I was stunned to see her. "What're you doing here?"

"Someone called me and said you were hurt. They told me to come right away. The receptionist would be gone, but I could use a code to get to the ninth floor." She showed Ian and me the number. "I've been knocking on every door trying to find you."

"Who told you that?" Ian asked.

"Someone from Harvester security," she said, her eyes wide with panic as she looked to Ian and me for answers.

Ian turned pale. "She's been very quiet," he whispered. He approached the birdcage and tapped it. "She's dead."

Sophie and I moved closer to look. The lifeless body of Ian's bird lay at the bottom of the cage.

Maru appeared at Ian's door, entered, and locked it behind him. "Good evening. Am I interrupting something important?"

"Get out of my office!" Ian yelled. "You filthy piece of shit! What did you do to my bird?" His voice trembled with grief and fury.

"What are you guys talking about?" Maru strolled over to a chair and sat down. "Narin, how are you? It's impressive how quickly you've climbed the corporate ladder here at Harvester. Are you enjoying the comforts of the executive suite, or do you just enjoy the company? Ian can be quite charming. He had me under his spell for a long time, and apparently, he's enchanted you as well."

"There's nothing you can do, Maru," Ian sputtered. "Everything has been set in motion. If you blow things up, you'll end up in jail for the rest of your miserable life." Ian swallowed hard and looked at his bird lying at the bottom of the cage. "But I'm willing to negotiate. We can work out something you can live with."

"What are they talking about, Narin?" Sophie asked. "What the hell is going on?" Her eyes begged me to explain this nonsense.

"Something I can live with?" Maru wiped his glasses. "I've been thinking about what I can and cannot live with for a long time. But before we go any further, let me clarify the situation for you. We knew from the beginning what you and your father planned. We have proof that you've been secretly giving Juan Alvarez proprietary information, which has allowed him to keep his lawsuit against Harvester going. You thought that crashing Harvester's stock would trigger calls for leadership change, giving you and your father the chance to take control."

He wiped his glasses again and continued, "We've already informed all the board members about your betrayal. Your father's proposal will be rejected at tomorrow's shareholder meeting before it even gets a vote. The inevitable outcome will be your dismissal and your father's removal from the board. The simplicity is elegant. Sadly, for you, the simple solution isn't the most satisfying. And I crave satisfaction." He pulled out a gun.

Holy fucking shit.

"They taught us in business school that most problems have a simple solution," Maru said. "But sometimes, the simple solution doesn't address the core demands of the problem. While it may require less effort to implement, it will ultimately cost you. So, what is the real problem we face?"

How long have you been practicing those lines?

"Please take a seat," Maru said. "We have a long night ahead of us."

We all exchanged glances, but after a brief hesitation, Sophie and I sat next to each other while Ian sat on a chair facing us. Maru remained seated, aiming his gun at Ian. Ian's hands trembled in his lap. I had never seen him like that before. In every situation, he had always been in control. His hands quivered like those of the cursed grandfather clock.

Maru pressed on. "The problem, as I see it, is that there are schemers among us," he said. "Their schemes can easily be thwarted, but justice requires so much more."

I watched Maru's eyes, his hands, and the gun. I had the knife in my right breast pocket. One opening, and I could jam it in his throat.

Beside me, Sophie let out a quiet moan. She wasn't looking at Maru; her eyes were fixed on Ian, who looked like a wounded, beautiful animal. Her gaze was full of tenderness and concern. She'd never looked at me that way. Maybe if Sophie had looked at me like that, even once, things might have turned out differently.

"Sophie?" Maru said. Sophie flinched, snapping out of her reverie. "What are you thinking about, Sophie?"

"Leave her alone," Ian croaked.

"Ian, I need to know the truth." Sophie looked at Ian with deep affection. "When I saw you at Narin's house, you seemed very familiar—too familiar. I didn't understand why until last night." She paused and briefly looked in my direction. I saw what looked like pity in her eyes. She continued, "Last night, I was going through old photos of Ravi and me and saw you standing in the shadows. You were looking right at me."

Ian's shaking worsened.

I realize now that you've been there all along, waiting in the shadows," she said. "You were the one who paid off my creditors in the past. I wondered why they'd stopped calling and harassing me. When my dad died, you stepped right up and were there for me. I appreciate everything you've done more than you'll ever know. And you did it all because you cared so much about Ravi."

Maru snickered, but she kept going, driven by her need to tell Ian how amazing he was. "Even after he died, you kept doing what he would've done—looking after me, taking care of me, helping me when I needed it most." She paused to catch her breath. "But I don't get why you were hiding, Ian. I would've loved to be your friend too. Why didn't you introduce yourself? The three of us— you, me, and Ravi— could have been the best of friends."

Maru chuckled. "This is perfect. It's exactly what I hoped for. This vain little brat thinks you wanted to be her friend. But it's a strange way to do that — stalking her and Ravi." He looked at me. "Does that make any sense to you, Narin?"

I shook my head.

Maru fixed his hateful gaze on Ian. "Why were you hiding in the shadows? Who were you looking at? Who were you stalking? Tell her the truth. It's about time you did."

"This isn't necessary," Ian whispered. "You've already won. Harvester is yours. I unconditionally surrender. What else do you want?"

"I want to hurt you," Maru said, his voice icy cold. He leaned in closer. "Tell them, Ian. Tell them the truth." He waved the gun at Sophie and me.

I asked, "Have you ever heard that violence is the last refuge of the incompetent?" I wanted to see if that would throw him off his game. It certainly did.

He hesitated and then aimed the gun at me. "Say that again," Maru yelled. "I dare you to say that again."

It would have been so easy. It would have been so fucking easy to take down Maru right then and there. But I didn't move. I just let things unfold. Team Chandra was on the rise, and I needed to be on the winning side.

"It was my father's idea, not mine," Ian said. "I didn't even want this."

"More lies," Maru shouted, leaping from his chair and knocking it over. "And we were so close to uncovering the truth. If you don't tell them, Ian, then I will."

Ian crossed his arms, trying to steady himself and stop trembling. I felt a little pity for him, but mostly I felt hatred. It was Ian's own damn fault that he was in this mess. He had lost the game. And Maru understood that letting the loser go free is a sign of weakness. At that moment, Maru wasn't weak at all. I was impressed that he had gone this far.

Maru began to pace. "Ian and Ravi had been quite the couple, prancing about town, spending nights locked away in this very room."

"Ravi?" Sophie's face twisted as she tried to understand. "I—I don't get it. Ravi and I were engaged. We were going to get married."

Maru shook his head and laughed. "Ravi and Ian were an item before you came along and stole him. Ian put on a good show after being dumped. He tried to make it seem like it was a mutual decision and that they stayed close friends after the breakup. But the facts tell a different story. Ian did everything he could to win Ravi back. He showered him with money and gifts despite their separation. He bought Ravi's paintings at exorbitant prices. Still, it wasn't enough. Ravi had set his sights on you, and nothing was going to stop him."

Sophie's face showed her confusion as she tried to understand what Maru was saying.

Maru continued, "I'd see Ravi every Thursday at noon and hand him his Red Sky for the week, a bottle containing a hundred red pills. He'd pay me with a wad of cash. Two years ago, I'd had enough. That's when I came up with a brilliant plan. With one blow, I brought the hand of vengeance upon three of the worst people in the world: Ian, Ravi, and Sophie. It was quite elegant if you think about it."

"So, you killed Ravi with tainted Red Sky," Sophie yelled, her voice laced with hatred. Her face, arms, and neck were bright red.

Maru laughed a fake laugh and kept going. "Fast forward, and here comes Narin, Ravi 2.0; similar looks, inferior in some respects, but more than enough as a replacement. This was a chance to start fresh and revive old passions. It was almost a perfect redo for both of you. You both could love him. Ian could build a friendship with Narin, and the three of you could live happily ever after."

"Maru," Ian whispered. "Shoot me now. Please. If you're going to do it, then do it now. Just let them go. That's all I ask. They're innocent."

Maru slipped behind Ian. The path was clear. I could have ended this, but I stayed still. I watched the scene unfold with a strange fascination. The game was over; Ian and Harold had lost. Maru was about to make his victory complete.

"You're the boss." Maru aimed the gun at Ian's right ear and fired. Ian collapsed to the ground.

Sophie covered her mouth and stared in horror at the lifeless body.

"I hope he meant that literally," Maru said. "Maybe he was being poetic. Ian tended to do that."

Sophie was gasping, sweat running down her face. "Why did you kill my dad?" she whispered. "I know only an evil prick like you would do such a thing."

Maru stepped over Ian and moved behind Sophie. He gently touched her face and aimed the gun at her temple. *Another chance to end this. Even Maru was probably wondering if I would at least try to save Sophie.*

"Why did I kill Morey? I'll tell you why." Maru's face contorted with hatred. "To punish you," he whispered in her ear. "Look over at Narin."

She turned to look at me. The redness had faded, and she looked as pale as a ghost. Tears welled up in her eyes.

"Why are you crying?" Maru asked. "You should tell him the truth. You should tell him that you never really loved him. You've just been using him to try to get Ravi back."

"You're as repulsive as ever," Sophie said.

"That's a terrible thing to say, Sophie," Maru whispered. "I was about to give you a high compliment. You're the only one of us with a pure heart. You've only loved one man all along. When you're with Narin, all you do is think about Ravi. Narin's just the living version, your second chance, the stand-in."

"That's not true," she yelled. "You worthless piece of shit. Morey didn't do anything to you! He spent his whole life building Harvester but didn't get any credit for his hard work. Your father took all the credit. You grew up in a million-dollar house because my dad worked his ass off to make Harvester successful. He didn't deserve this."

Maru whispered into her ear, "Do you mean your father, who wept at his desk with no one to comfort him? Do you want to know how many nights I listened to him moaning, wretched and drunk, crying because his selfish little brat of a daughter abandoned him? You've always been an arrogant bitch, ever since you were a kid."

"And you've always been a weaselly little piece of shit," she said coolly. "And you still are."

"Is that so?" Maru moved behind me but kept the gun pressed to Sophie's temple.

Last chance. It's now or never.

"Now, Narin, my friend, what am I supposed to do with you?" he said. "You see, these white folks, they always abuse us. They've always done it. They always will. No, no, no. Don't you dare do that, or I will shoot her."

He probably thought I was about to attack, but that was the last thing on my mind. I was ready to tell him I was fully committed to Team Chandra. I was just waiting for him to finish with Sophie so she wouldn't find out about my treachery.

"And—And—You!" Maru shook with rage. "You came to Harvester and immediately started plotting with Ian to take everything away—take away everything my father worked his whole life to build. You decided you had the right to throw my father out like he was garbage. I gave you a chance to redeem yourself, and once again, you chose the wrong side. Tell me, what option do I have left? What does justice demand I do to you?"

I looked at Sophie one last time. Her eyes had changed; they were empty and hollow, no longer full of life. I felt a needle pierce my neck. Everything turned red and then black.

I woke up on the floor next to where I'd been sitting. Pain shot down my right arm as I pushed myself up. My head spun, and the room whirled, forcing me to collapse back onto the floor. I lifted my heavy head and shook it to clear the dizziness. My fingers sank into the plush red velvet carpet, and my muscles ached as if I had been convulsing for days. I shook my head again, trying to remember how I ended up on the floor.

"Are you awake?" Maru asked. "You should get up now. There's not much time left."

I lifted my head and saw Maru looming over me. I remembered that Maru was my friend. I needed to tell him something, but I couldn't recall what it was. He reached down to help me up. With some effort, I stood on my feet.

"Something bad happened while you were asleep," Maru said. "Take a look."

I followed Maru's gaze. Ian was lying face down on the floor, blood gathering around his head. Sophie sat on the couch, her eyes wide open, unblinking and unseeing. Her dress was stained a deep crimson.

"Does that make you sad?" Maru asked.

I shook my head. I wasn't sad at all; I was just relieved that it was finally over.

"Go give Sophie a goodbye kiss."

I followed his instructions and kissed her cold forehead. She carried the scents of moss, rain, and flowers. I lingered for a moment, captivated by the aroma.

"Are you feeling okay, Narin?" Maru asked. "You don't look well."

I started to remember what had happened. He must have injected me with DMTA and H138, just like he did to Morey. I looked at Sophie's painting. The mangoes were a swirl of bright yellow, orange, and red, like a sunset spiraling down a drain. The painting spun like a vortex, threatening to pull me in. The feeling was both thrilling and terrifying.

"You all were taking Red Sky together, having a great time, but then you felt the switch coming on," Maru said. "They tried to stop you, but you pulled Ian's gun out of his desk and shot both of them."

I nodded.

"Did you know Ian had a loaded gun in his desk?"

"No," I replied.

"Now you do." Maru smiled. "With them gone, you had to finish what you started."

I nodded again. So, that was his plan: to kill everyone and make it look like I did it.

"We only have about five minutes before the switch," he said. "Sit down. I've got a few questions to ask you."

We sat on the ground, facing each other, with the gun on the floor between us. Maru kept glancing at his watch.

"First question: Why did you choose to work at Harvester?"

The interrogation had started, and I couldn't hold back. I couldn't lie or tell him to go screw himself. I couldn't even lie to myself.

"Because Ian and Morey convinced me we could develop a revolutionary new drug that harnessed the beneficial effects of Red Sky without concerns about addiction and suicide."

"But why was that important to you?"

"I wanted to be a great man," I said. "I wanted to be famous and respected so that people would like me, be interested in what I had to say, and genuinely want to be my friend."

"And you wanted to help the world?" he asked. "Is that it?"

"Sure," I said, shrugging. "Mostly, I wanted to help myself."

"By becoming famous?"

I shook my head. "You see, Maru, I've never had any real friends. I've always been alone and deeply depressed." He visibly winced when I said this, but I kept going. "I've been hospitalized twice for severe depression and suicidal thoughts. But it's not just that. I've always blamed myself for not being as smart as I could be. My memory isn't great compared to some of the brilliant people I've met. Don't you see, Maru? I wanted a pill to make me better than I am. I hoped to be the first of a race of superhumans with unlimited intelligence and no emotional baggage. I saw it so clearly. I dreamt of the day I'd become the first perfect human being. Then people would finally care about me and want me in their lives."

He nodded at me sympathetically. "Good answer," he said. "Second question: Did you love Sophie?"

I had to pause for a moment to think. "I thought I did," I said. "I thought she was all I needed, but then I realized she wasn't."

"Was it because of what you saw in the red room?"

"It was a lot of things," I said. "I mean, she's an artist, for God's sake. What kind of profession is that? And she's an alcoholic. And she doesn't even listen to me half the time. And—And the way she clung to Ian at Morey's funeral. You saw that yourself. I mean, what kind of perfect girl does something like that? No, she wasn't the right girl for me. She didn't understand me. I'm glad she's gone."

Maru replied, "That's quite harsh."

"Every time I thought about telling my parents about Sophie, I'd panic. On my way back from Chicago, I decided I had to cut her out of my life. And you did that for me. I knew I was free the second I saw you pull out the gun. I guess I should thank you for that."

"You let me kill her so you could be rid of her?" He looked genuinely surprised.

"That's right," I said. I struggled to find a way to express my complete disregard for the lives of those closest to me. "I'm a terrible person, Maru," I said. "I'm selfish as all fuck."

Maru nodded and glanced at his watch.

"We have time for one last question. Why did you go all in with Ian instead of trying to be friends with me? We could have been such great friends, you and me. I could have been the best friend you ever had."

When I didn't answer right away, he kept talking. "I've never had friends either. My whole life has been one lonely nightmare. But Harvester took it to the next level. Everyone at Harvester hates me." He started to choke up. "But it would have been great to have at least one friend, someone I could talk to, someone who didn't hate my guts. But—But then you chose Ian over me." Tears streamed down his face. He took off his glasses and wiped his wet cheeks with his sleeve. "I guess I'm not surprised. But still, I don't understand what I have done to deserve all this hatred?"

"I know why everyone hates you," I said.

He looked up at me and asked, "Why is that?"

"Whenever you try to carry out your plans, your father informs the leadership that they can ignore them." I paused to let that sink in. "Everyone knows you lack any real power because your dad won't support you. He undermines you behind your back."

"Who told you that?" he croaked.

"Ian told me, but everyone knows, even the janitors," I said. "You're the only one at Harvester who doesn't see that your dad has no faith in your judgment. Everyone thinks you're a fly buzzing in their ear when they're trying to focus on their work. They'll shoo you away if they're feeling nice or swat you if they aren't. But no one will love or respect you when you have nothing to offer in return."

"My father does that?"

I nodded. "I can't lie to you with that shit you injected into me. I wish I could, but I can't. I feel sorry for you. I really do."

He took several deep breaths. "Thank you, Narin, for sharing that with me." He looked so distraught that I felt an urge to jump up and hug him.

"I also know why you hated me from the very start," I said.

"I didn't hate you," he said defensively. "Why would I hate you?"

"Because I remind you of Ravi, and Ravi was everything you are not. He was talented and successful. Ian and Sophie found him irresistible. But you're just an untalented failure."

"That's not a very nice thing to say," Maru responded.

I said with a warm smile, "I've got to tell the truth."

We sat there, staring at the gun. He picked it up, wiped it with a towel, set it down again, and then stood up.

"Goodbye, Narin," Maru said. "I'd say it was a pleasure knowing you, but I can't. You see, I also have to tell the truth."

Without looking back, he walked out the door and shut it behind him.

I sat there, staring at the gun. The knife burned in my coat pocket, so I pulled it out and held it up. It glowed so brightly it hurt my eyes. It was on fire. I felt my hand burn. I could smell the burning flesh. I thought about stabbing the knife into my chest, but I didn't. It would have been so easy, yet I just couldn't do it. The pain became so intense that I dropped the knife and collapsed. My muscles trembled, but I couldn't move. I had lost control of my body. Bright lights soared high above me, sparkling and exploding in the sky.

I lay beside Ian on the boat. Ian told me how bored he was with life. He said that everything he did was meant to stave off this terrible boredom. Each precious day, everyone fights that wicked demon, that pernicious snake. But boredom always slithers nearby, ready to strike. It sleeps lightly and awakens when the explosions and glittering lights turn to ash. That's why we do the terrible things we do. That's why we drink, gamble, and pretend to be prettier and smarter than we are. That's why we get up every morning, work hard all our lives, or take a fool's gamble so we might catapult ourselves past the fangs of that wicked snake. We do all this to keep the snake in our souls quiet for a moment longer, so that, for one more turn, we can avoid its gaping mouth.

I lay beside Sophie as she told a story about an old woman she met on a dirt road at dusk. The woman wore a necklace made of skulls, and Sophie rubbed the shiny skulls between her fingers, admiring their beauty. The old woman gently touched Sophie's face, and Sophie ran her fingers through the woman's long, silver hair. The sky darkened, and a cold wind blew. A black bird soared above the grassy field as they walked along the dirt road together. The sun set on the horizon, and they passed an ancient, dilapidated gate before entering the dark forest.

Sitting at the kitchen table with my grandmother, I listened to the scratchy sounds of a song playing on the record player. But now I understood the meaning without translation: Oh Mother, how I long to be with you. A moment without you is like an eternity of suffering. Please take me now so that I can enjoy your magnificent bliss. This joyless life has nothing to offer me.

I hadn't understood it before, but now I did. The mournful singer wasn't singing about death; he was singing about life. Living begins when we stop perceiving separation. There is no separation except what we create in our minds. All things are interconnected. I never realized this until I had Red Sky injected into me.

I crouched inside a bright red carriage with gilt trim, racing at full speed. I raised my head and was hit with a blast of hot wind and sharp sand needles. Before me stood the Desert Prince, his robes billowing wildly behind him. He stood at the chariot's helm, his face stoic as he gazed forward, seemingly unaware of the swirling sand and fierce winds. Two massive red oxen pulled the chariot at breakneck speed across the desert.

An overwhelming wave of guilt flooded over me. I knew I had let him down. He probably saw me as a coward. I dragged myself across the noisy floorboards and knelt at his feet. I couldn't find the right words to express my regret. I stayed there, face down, with my arms stretched out in front of me. That's when I noticed that my arms were covered with fine, downy hair. No, not hair — it was fur. I looked down at my torso and saw I was covered in a tawny pelt. My hands and feet had tough, leathery skin. I tried to say something, but only a grunt escaped.

The Desert Prince spoke. "It was foolish of me to try to guide you through life's pitfalls. My training, meant to empower you, only made you culpable. Had I not prepared you for battle, you would have remained naive, weak, and frightened, but you wouldn't have become an active participant in your own destruction. As you gained strength, you became vulnerable to moral decay. With my rising expectations, you embraced your basest instincts."

He continued, "You defeated Monkey, but you failed to achieve his nobility. Now you see for yourself what you have become. You have become pitiful and have degraded yourself beyond measure. You wish to be the perfect man, but there is nothing more vile than strength that is not tempered by love and compassion. What I should have understood is that your self-absorption and ego would triumph over love, friendship, and family. I now realize that my teacher's lineage is coming to an end, and his wisdom and nobility will forever be lost to this world."

The chariot raced recklessly toward the cliff's edge. In that moment, I realized no one was holding the reins or steering it. I was just along for the ride. Then, I plunged off the edge.

Sounds echoed nearby, causing my muscles to tighten. I strained to turn my head. The hallway buzzed with voices and jingling keys. Strong arms grabbed me and lifted me. My vision blurred. The last thing I saw was five mangoes, stacked precariously on top of each other, ready to topple over.

When I woke up, I found myself here in this jail cell, awaiting my fate.

Epilogue

You're probably wondering what happened to me after all that. Now it's 2007. It feels like April or May, but I don't fuss over tiny details anymore. I decided to write a follow-up to my confession to show you that everything has turned out well for me. In fact, I'm happier than I ever thought I could be.

To sum up, on July 17, 2006, Maru planned to kill Ian, Sophie, and me. I barely survived, but Maru made it look like I was responsible for the murders.

I've tried to piece together what led to Maru's killing spree. Here's my theory: Maru murdered Ravi out of jealousy by using contaminated Red Sky. It's unclear whether Maru and Ian had a sexual relationship before Ian met Ravi. Whatever their relationship was, it was more than just friendship or being frenemies.

Maru felt betrayed when Ian got involved with Ravi. In killing Ravi, he also seriously hurt Sophie, but he didn't seem too bothered by that, given their troubled history (remember, she had dissed him when he asked her to the prep school prom). Ian was devastated by Ravi's death and felt partly responsible for Sophie's injuries. What's unclear to me is whether Ian's or Maru's version is correct. Ian and Ravi might have had a romantic fling, but they stayed friends after Ravi started dating Sophie. If you believe Maru, then Ian never accepted the breakup and was obsessed with winning Ravi back. The truth might be somewhere in the middle. We'll never know since both of them are dead.

Regardless, Ian became obsessed with Sophie. As a result, he spent his last two years watching over her, secretly paying her bills, and caring for her. I had no idea how he knew so much about her, but now the answer seems obvious. He had a private detective keep tabs on her. If I'd thought of that before, I could have been on the lookout for a suspicious-looking man following us around. Oh well.

Sophie had always rejected my offers of help, but on her last day, did you notice how she gushed about how thankful she was to Ian and Ravi for "taking care of her"? That really pissed me off. But what's done is done.

When I first entered the picture, Ian had the bright idea to set me up with Sophie. I could serve as a perfect substitute for Ravi. After all, he was my older and more talented twin brother. Ian also liked me for the same reason: I reminded him of Ravi. So, Ian and Sophie redirected all their affection for Ravi onto me, which worked well for a while until it didn't.

From the start, one burning question was why Maru hated me so much. Now, it's clear that I also reminded him of Ravi. Remember, he killed Ravi out of jealousy. Mystery solved.

The more complicated question was whether Ian secretly loved Sophie. I believe Ian cared for her, but he wasn't interested in a relationship. He had many guys and girls to choose from, so I couldn't see him settling on just one girl.

But why did he keep the paintings in the Red Room? Maybe he had feelings for Sophie, or perhaps they reminded him of Ravi. After all, they were his paintings. I was too close to the situation to be objective. I tend to overthink things and let jealousy take over, so I'm not the best judge. Maybe, if events had followed their natural course, Ian and Sophie would have fallen for each other and left me behind. It definitely seems like that's where things might have ended up if Maru hadn't interfered. Again, we'll never know since the principal characters are dead.

The biggest question is why I didn't stop Maru when I had the chance. With my training and the knife ready, I could have easily taken him down.

That's exactly what the Desert Prince thought I'd do. What he didn't realize was how selfish I was. I jumped on the Team Chandra train when I saw the gun. The Chandras were going to win, so why stay on a train that's about to derail? If I had the chance, I would've told Maru that, but I was waiting for him to shoot Sophie first. For some stupid reason, I didn't want her to hate me, even if she was about to die.

Maru messed things up by injecting me before I could explain myself. Then, when he questioned me, I insulted him instead of trying to win him over. You can't really lie with Red Sky on board. Well, what's done is done. It's a lot like Hamlet. Everyone's dead, so there won't be another act.

In the early days, while sitting in prison, I wondered if something might happen to Maru. Maybe he'd feel remorse and take his own life. If I ever got out, I thought I could take his place in Sanjay Chandra's eyes. If I can be a stand-in for someone with real talent like Ravi, I could be a good substitute for a loser like Maru.

You might also wonder why I didn't go into the switch with DMTA and H138 on board. Maru expected I would kill myself after he left me with the gun right in front of me. Well, it turned out that since Morey's funeral, I'd been injecting myself with H140, the Type II receptor blocker. I had no idea how to dose it, so I used an equivalent dose based on the amount we were giving to rats according to my weight.

I didn't even know if it would work or if it would kill me. It hadn't been tested on humans before. But I was so paranoid that what happened to Morey might happen to me that I felt I had no choice. I guess it wasn't paranoia, since it turned out to be true. That was a smart move on my part.

A lot has changed since that day. The Red Sky left my system, and I slipped back into the dull, monochrome "normal" world, filled with chaotic thoughts and stifling inaction.

On the third day of my detention, Officer Coffin told me I was charged with the murders of Ian Blair and Sophie Whitely, and that I would stay in custody until my bail hearing in a week. He said that my parents had hired a lawyer and were traveling from Chicago. My migraines and depression came back stronger than ever.

When my parents arrived, I felt really sorry for them. My mom looked awful, as if she had aged ten years. I guess having your son arrested for murder can really wear someone down. They told me they had hired the best lawyer and that I would be out very soon.

Dad told me he'd heard from a family friend that I could get my pharmacy license in less than a year with some extra training. I could stay at home and work at the Walgreens down the street. I tried to smile, but I couldn't. I thought about how Deepa hadn't come with them. She couldn't get past her hatred, even to visit me in my time of need.

The lawyer tried to help me, but at the bail hearing, I told the judge I didn't want to be released. This shocked my parents, and my lawyer yelled at me, accusing me of being uncooperative. Still, I had my reasons. I went back to jail, a higher-security facility than Coffin's jail, and waited for my trial.

At first, Sanquel visited frequently, but he soon realized I didn't need him to "lift my spirits." The final straw was when he handed me a copy of Professor Goldblum's obituary, which I barely glanced at. After that, he didn't come back.

Floyd visited me once and told me he was the one who left the note warning me about Ian: 'Don't believe anything he says.' He tried to help me, and I appreciated his effort. Floyd also mentioned that Maru had killed himself. "Good," was all I could say. By then, I had no interest in returning to my previous world of test tubes and research. I asked Floyd for a favor, and he agreed to it.

None of the police investigators showed any interest in my "confession," and I never shared it with my lawyer. I kept it hidden under my bed. I didn't write anything else while I was in jail; I found a better way to pass the time.

Once Red Sky was out of my system, I started craving it again. I missed the incredible clarity it gave me. In jail, it was easy to find someone who could sell Red Sky at a fair price. I took it every day, and I felt like I was in heaven. I know I could have gotten Red Sky on the outside, but then I'd have to make new contacts, and I wasn't in the mood to meet new people.

One day in October 2006, a police officer stopped by and told me that the charges had been dropped. He didn't explain why. My lawyer must have found something that cleared my name. I was surprised; I thought they had a pretty solid case against me.

They returned my belongings: my wallet, clothes, knife, and Ian's Porsche keys. I didn't go back to the car because I wasn't sure it would still be there, and I didn't want to go anywhere near Harvester.

The first thing I did was drain my bank account. I had a few thousand dollars left from my time at Harvester. I also tried to make as many cash withdrawals as possible on my credit cards before they got canceled.

I went downtown and spent my days walking along Market Street. I'd sleep on the sidewalk or under a bridge, wherever I could find a spot. I would sit on the steps of the Peabody Opera House, watching the Red Skyers pass by. They were easy to spot and always super friendly when I told them what I wanted. They reminded me of evangelicals when you tell them that you want to know more about Jesus. They'd take me into the alleyways where I could buy as much Red Sky as I wanted. I've since discovered other ways to get Red Sky without using any money. It's disgusting, but it's worth it.

I met Floyd around that time. He was the last person I called before I threw my phone into the river. I gave him a hundred bucks, and he handed me the painting of the five mangoes. He said it had gone into storage at Harvester, but he was able to find it.

Eventually, I found a small, run-down apartment and rented it under a false name so no one could track me down. The place I'm staying is infested with cockroaches and rats. My former self would have hated it, but I no longer care about such things.

My family has no idea where I am, which is probably for the best. They'd be shocked by how I look. I've stopped shaving and showering. I don't have a single clean piece of clothing. These superficial extras are part of the limited reality most people experience. I don't need any of that. I barely eat or sleep now. Red Sky provides me with everything I need.

I've pulled out most of my hair. I doubt anyone from my past would even recognize me. I used to pull at my hair when I was stressed, but now I do it because it gives me intense pleasure.

I'm sitting on my bed, staring at the painting of five mangoes. It tried to pull me into its vortex the day all that shit happened. I thought it was trying to swallow me. But now I see it was trying to save me.

I take a pill every few hours and stare at the painting day and night. The colors glow and dance, pulling me into a new world where dark clouds swirl beneath a blood-red sky. It's impossible to describe this place. There's no war, terrorism, torture, or crime. Everything pulses with life and fills me with immense joy. The beauty of this world makes me want to cry. Tears are welling up just thinking about it.

I'm happily married in this wonderful world. Each day, I have a different wife—an endless series of beautiful women I can enjoy whenever I wish. The sex is incredible. Despite their superficial differences, all my wives share the same qualities: they are virgins who adore my parents and would do anything to make me happy. They're all highly educated; some are doctors, while others are engineers, and they don't worry about anything at all.

Whenever I call my wife, she drops everything to see how she can make me happy. She looks at me with so much love, warmth, and understanding that sometimes I can hardly bear it. Nothing matters more to her than making me happy in every way.

My parents tell all their friends about their amazing, talented daughter-in-law who loves them wholeheartedly.

We have two wonderful sons, *Bakasura* and *Kirmira*. We chose Indian names to please my parents. Although they can be a handful sometimes, we love them dearly.

Every week, we host a gala at our mansion where accomplished individuals gather to discuss art, literature, philosophy, and science. Everyone in the jet-setting community longs for an invitation to our parties. My wife is a charming hostess who thrives on the admiration of our wealthy and famous friends. Everyone she meets is captivated by her beauty, elegance, and charm.

I have so many friends that I can't keep track of them all. I've won the Nobel Prize for my work, and everyone praises my brilliance. They all want to talk to me and listen to everything I say. They laugh at my jokes and admire my cleverness. Every picture we take together turns out perfectly, showcasing our ideal life to the world.

As you can see, I'm in heaven. I have everything I ever wanted. Everything is within my reach.

I'm sorry to say this, but your reality is merely an empty shell, a shadow, a nothing compared to mine. I pity you for wasting your days in such a meaningless way.

I reflect on my past and realize how self-absorbed and arrogant I was. My goal was to "tame" Red Sky, to take away its control over life and death. What a blind fool I was. I didn't see that Red Sky is the answer. It's everything: pure, unfiltered Red Sky, without any extras. Red Sky has given me more than all my years of schooling, tens of thousands of dollars in therapy, my family's love, and all the so-called love and friendship I had before my transformation. Red Sky is everything; all else pales in comparison.

You might think I'm foolish for taking Red Sky while I'm alone in my run-down apartment. I understand that the switch will happen someday, and no one will be here to help me. I'm okay with that. It could happen today or in three years. I don't care.

I've opened my eyes and seen the light. Red Sky is the Eternal Harvester, and we're just grain waiting to be gathered. How can a grain resist the incredible power of a steel-bladed thresher?

My sharpened knife sits at my bedside. The serrated edges burn like fire.

My knife is ready, and I am ready.

I am waiting for Red Sky to call me to my eternal home.

This is the first book in the Juggernaut Trilogy. As advances in biotechnology, artificial intelligence, and medicine accelerate, the trilogy asks: What happens when our technological prowess grants us powers before we have the wisdom and ethical framework to wield them for humanity's benefit? Find out more on my website.

ABAcharya.com

About the Author

My wife and I are both neurologists living in St. Louis, Missouri. I feel very fortunate to be married to my best friend, who is also an excellent editor and proofreader. My fiction explores ambition, power, and the unforeseen consequences of human ingenuity. If you want to learn more about why I wrote *Red Sky*, you can find essays and reflections on my Substack.

aba55.subtack.com

If you enjoyed this book, I'd appreciate it if you left
an honest review using the link below.

You can also find the book on Amazon.com at
https://www.amazon.com/dp/B0GJJGXLTX

Please join my mailing list using the link below,
where I discuss themes from *Red Sky*, run monthly giveaways,
and offer early access to forthcoming books.

www.ingramcontent.com/pod-product-compliance
Lightning Source LLC
Chambersburg PA
CBHW070516140726
48132CB00030B/391

9798995219606